simple methods for aquaculture

MANAGEMENT

for freshwater fish culture

ponds and water practices

Text: A.G. Coche
J.F. Muir
T. Laughlin
Illustrations, book design
and layout: T. Laughlin
Artwork: E. D'Antoni
Production: J. Lore

FOOD AND AGRICULTURE ORGANIZATION OF THE UNITED NATIONS
Rome 1996

David Lubin Memorial Library Cataloguing-in-Publication Data

FAO, Rome (Italy)

    Management for freshwater fish culture:
     ponds and water practices
    (FAO Training Series, No. 21/1)
    ISBN 92-5-102873-7

    1. Fish culture   2. Fish ponds
    I. Title II. Series

    FAO code: 44   AGRIS: M12

THE AQUACULTURE TRAINING MANUALS

The training manuals on simple methods for aquaculture published in the FAO Training Series are prepared by the Inland Water Resources and Aquaculture Service of the Fishery Resources and Environment Division, Fisheries Department. They are written in simple language and present methods and equipment useful not only for those responsible for field projects and aquaculture extension in developing countries but also for use in aquaculture training centres.

They concentrate on most aspects of semi-intensive fish culture in fresh waters, from selection of the site and building of the fish farm to the raising and final harvesting of the fish.

FAO would like to have readers' reactions to these manuals. Comments, criticism and opinions, as well as contributions, will help to improve future editions. Please send them to the Senior Fishery Resources Officer (Aquaculture/Publications), FAO/FIRI, Viale delle Terme di Caracalla, 00100 Rome, Italy.

The following manuals on simple methods of aquaculture have been published in the FAO Training Series:

Volume 4 — Water for freshwater fish culture

Volume 6 — Soil and freshwater fish culture

Volume 16/1 — Topography for freshwater fish culture: Topographical tools

Volume 16/2 — Topography for freshwater fish culture: Topographical surveys

Volume 20/1 — Pond construction for freshwater fish culture: Building earthen ponds

Volume 20/2 — Pond construction for freshwater fish culture: Pond-farm structures and layouts

Volume 21/1 — Management for freshwater fish culture: Ponds and water practices

A final volume is being prepared:

Volume 21/2 — Management for freshwater fish culture: Farms and fish stocks

HOW TO USE THIS MANUAL

The material in this manual is presented in sequence, beginning with basic definitions. The reader is then led step by step from the easiest instructions and most basic materials to the more difficult and finally the complex.

The most basic information is presented on white pages, while the more difficult material, which may not be of interest to all readers, is presented on pages with a grey or light blue background.

Some of the more technical words are marked with an asterisk (*) and are defined in the Glossary on page 231.

For more advanced readers who wish to know more about pond and fish management, a list of specialized books for further reading is suggested on page 233.

CONTENTS

2   IMPROVING POND WATER QUALITY

20 Introduction

1. Water is essential for the life of fish. It is the medium that must supply or support all their needs, including breathing, eating, reproducing and growing.

2. In a previous manual (**Water for freshwater fish culture**, *FAO Training Series*, **4**) you learned:

- where water comes from and where it goes;
- which water to use to fill your ponds; and
- how much water you will require.

3. Here you will learn a bit more **about water itself**, and about the main physical and chemical characteristics that are important for the production of fish in earthen ponds (see Sections 22 to 25).

4. Once you know your water well, you will understand more easily how to control its quality and, if necessary, improve it through good pond management practices (see Sections 26 to 29).

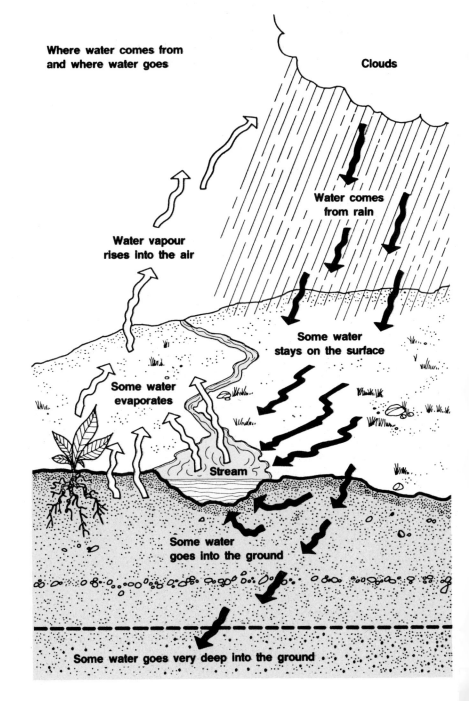

Where water comes from and where water goes

Clouds

Water comes from rain

Water vapour rises into the air

Some water evaporates

Some water stays on the surface

Stream

Some water goes into the ground

Some water goes very deep into the ground

Pond water composition

5. Pond water contains two major groups of substances as shown in the attached chart:

- **dissolved substances** made of gases, minerals and organic compounds;
- **suspended particles** made of non-living particles and very small plants and animals, **the plankton***.

6. The composition of pond water changes continuously, depending on climatic and seasonal changes, and on how a pond is used. It is the aim of good management to control the composition to yield the best conditions for the fish.

General composition of water

Dissolved substances

Gases
- oxygen, carbon dioxide, hydrogen sulphide

Minerals
- salts of calcium, magnesium, sodium, potassium, iron
- compounds of nitrogen, phosphorus

Organic compounds
- proteins, carbohydrates

Suspended particles

Non-living particles
- minerals such as silt and clay
- organic materials such as detritus, dead organisms, humus*

Microscopic living organisms
- plant forms (**phytoplankton***)
- animal forms (**zooplankton***)

7. Some of these substances are of particular importance for successful fish farming. You will therefore learn more about:

- **the particles suspended** in pond water (see Section 23);
- the living **plankton** (see Section 101, **Management, 21/2**);
- the **dissolved minerals and organic compounds** (see Chapter 6);
- **dissolved oxygen** in particular (see Section 25).

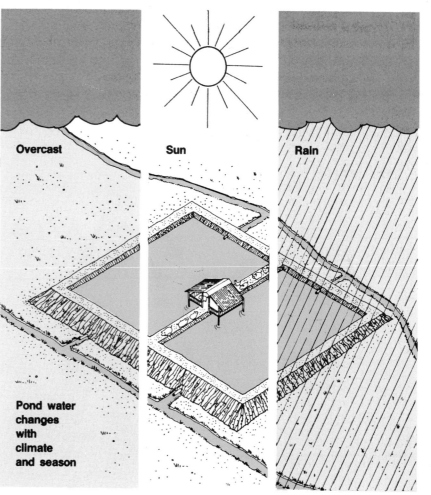

Overcast   **Sun**   **Rain**

Pond water changes with climate and season

Changing water composition

8. The characteristics of pond water depend both on the water that has been used to fill the pond and on the characteristics of the soil. However, within the pond water, some major chemical processes take place:

- **respiration*** (plants and animals): oxygen gas is consumed and another gas called **carbon dioxide** is produced;
- **photosynthesis*** (plants only): whenever sufficient light is available, carbon dioxide is used for the production of plant material, while oxygen gas is released from the plants;
- **decomposition***: dead plants and animals decay under the action of minute organisms called **bacteria***, and oxygen is used to produce mineral and organic compounds.

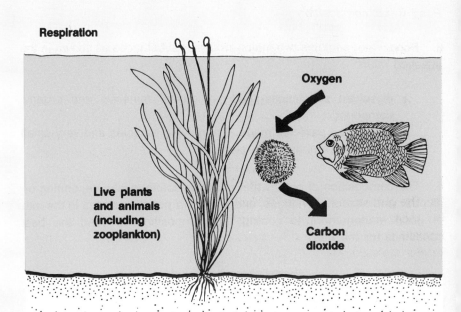

Respiration

Oxygen

Live plants and animals (including zooplankton)

Carbon dioxide

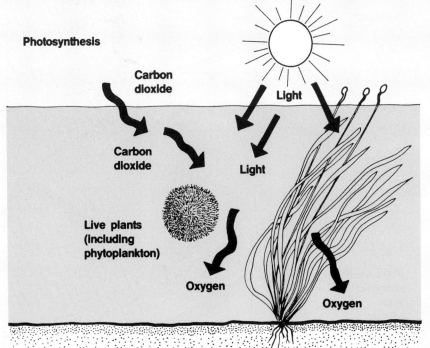

Photosynthesis

Carbon dioxide

Light

Carbon dioxide

Light

Live plants (including phytoplankton)

Oxygen

Oxygen

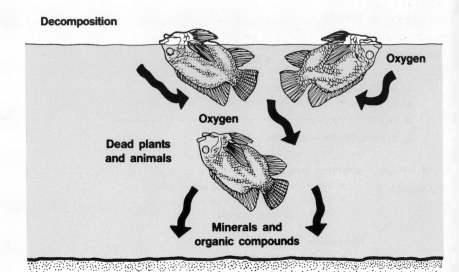

Decomposition

Oxygen

Dead plants and animals

Oxygen

Minerals and organic compounds

9. These processes constantly change the water composition, for example:

- **during the day**, by increasing the oxygen production and decreasing the carbon dioxide content through photosynthesis;
- **during the night**, by decreasing the oxygen content of the water and increasing the carbon dioxide content, through respiration in the absence of photosynthesis.

10. The greater the quantity of plants, animals and bacteria in the water, the more these processes change the water composition. In heavily stocked ponds, therefore, these changes are greater and need more careful management.

11. All of these chemical processes are influenced by **the water temperature**: the warmer the water, the more these processes increase, and the more quickly the water composition can change.

12. To manage and control the water composition, you need to **sample and measure** the composition, particularly of the more important characteristics. In the next sections you will also learn more about the four water characteristics that are of particular importance for fish pond management:

- **chemical reaction** of the water (pH);
- **turbidity**;
- **water temperature**; and
- **dissolved oxygen content**.

First we will look at how to sample the water so these factors can be measured.

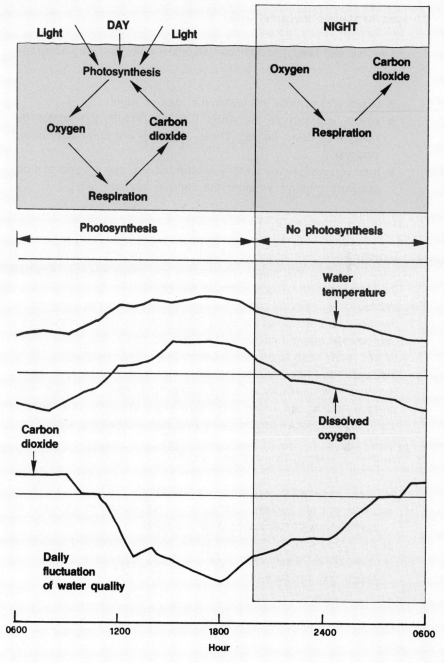

21 How to sample the pond water

1. As we will see later, the composition of the water can vary according to:

- **when** you sample the water (i.e. day or night);
- **where** you sample the water (i.e. the middle, the edges, the surface or down below). These locations are commonly called **stations**;
- **how** you sample the water (i.e. whether you use a basic or more accurate method, whether the sample is clean, etc.).

2. There are three ways in which water is usually sampled.

(a) **Directly**, with test materials or with an instrument. This method is best for getting immediate information and often lets you measure the water exactly where it lies in the pond.

(b) **Indirectly**, by using a bottle, bucket or other container, and testing the water at the side of the pond. This method may be necessary if you have to add chemicals to the water to test it. It is more difficult to get water from a precise location this way.

(c) **Indirectly**, as above, but **taking the water to a laboratory** to test it. Although the methods and the equipment used can give more precise results, the water may have to be specially preserved to make sure it does not change during transport to the laboratory.

3. Whichever method is used, you should:

- make sure all your equipment is clean;
- rinse out all buckets, bottles and instruments to be used for sampling with the water to be sampled;
- try not to disturb the water while you are sampling it; and
- note **when, where and how** you have made the sample and taken the measurement.

Obtaining a good water sample using a sample bottle

4. This procedure is the best for testing for dissolved oxygen using chemical methods (see Section 25). It may also be used for other analyses. Use **a narrow-mouthed glass bottle** of known volume, for example 100 ml or 250 ml. This sample bottle should first be washed out and rinsed with the pond water, then carefully filled, avoiding splashing or bubbling. Then with the bottle held below the water surface, push the stopper in. Take care to avoid trapping air bubbles in the neck of the bottle.

5. This method is good for surface and shallow waters. To obtain a sample from **deeper water**, however, you need to transform the sample bottle into a water sampler.

6. You can build **a simple water sampler** in the following way.

(a) Obtain a **narrow-mouthed bottle**, preferably made of glass and containing not more than 500 ml of water.

(b) Obtain a good **stopper** that fits tight into the mouth of the bottle.

(c) Firmly attach **a weight**, such as a stone or a heavy piece of metal, to the lower part of the bottle so that the bottle will easily sink.

(d) Attach **a piece of string** to the neck of the bottle, slightly longer than the maximum depth of the water to be sampled.

Note: instead of only one piece of string, you can also use two, one attached to the bottle and the other attached to the top of the stopper.

(e) Attach the stopper securely to the same piece of string, just above the opening of the bottle, **at a distance equal to at least twice the length of the stopper**.

(f) **Mark the string** with knots spaced at fixed intervals, such as 20 or 50 cm apart, so that you will know at which depth the bottle mouth opens to obtain the water sample.

A simple water sampler
made from a 250-ml bottle
with a tight stopper
and a weight attached
to the bottom
of the bottle

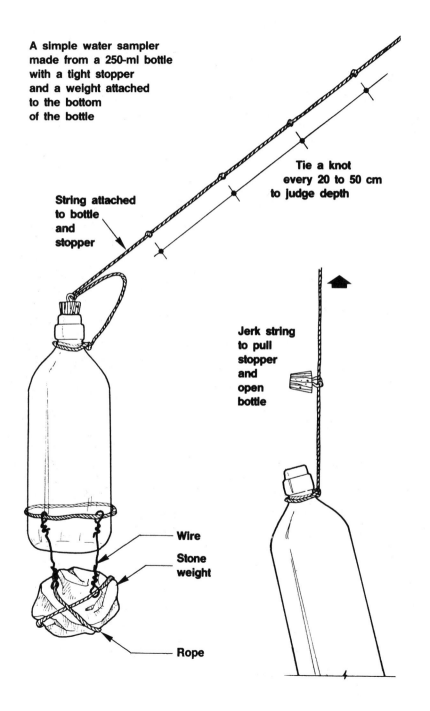

String attached
to bottle
and
stopper

Tie a knot
every 20 to 50 cm
to judge depth

Jerk string
to pull
stopper
and
open
bottle

Wire

Stone
weight

Rope

7. You may also secure the sampling bottle to **a piece of wood** with a rubber strap, for example, attaching a string to the top of the stopper, as shown in the drawing.

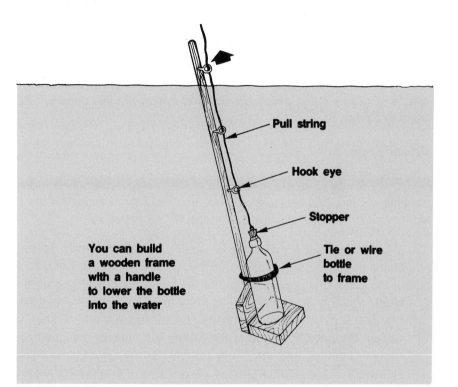

Pull string

Hook eye

Stopper

You can build
a wooden frame
with a handle
to lower the bottle
into the water

Tie or wire
bottle
to frame

8. **To obtain a water sample** from a certain depth, proceed as follows.

(a) Put the stopper firmly in place to close the water sampler.
(b) Lower the sampler into the water to the desired depth.
(c) Jerk the string sharply to pull the stopper out and open the sampler. It should now start filling with water, while air bubbles appear on the surface of the pond.
(d) When air bubbles no longer appear, carefully pull the full water sampler out of the pond.
(e) Immediately measure the water temperature (see Section 24) and the chemical characteristics.

9

22 The chemical reaction of the water (pH)

What does pH mean?

1. Water may be **acid**, **alkaline** or **neutral**. Depending on this, water will react in different ways with the substances dissolved in it. It will also affect in different ways the plants and animals living in the water. The measure of the alkalinity or acidity of water is expressed by its **pH value**. The pH value ranges from 0 to 14, with pH 7 indicating that the water is neutral. Values smaller than 7 indicate acidity and values greater than 7 indicate alkalinity (see Section 41, **Soil and freshwater fish culture**, *FAO Training Series*, **6**).

Measuring pH

2. Obtain a water sample, using one of the methods described in the previous section. You can use methods and tools to measure the pH of water similar to those you used to measure the pH of soil.

(a) **pH indicator paper**: a thin strip of paper (such as chemically treated litmus paper) is partly dipped into the water to be tested. The colour of the paper changes, and this new colour is compared to **a colour chart**, which gives the pH value according to the colour obtained. You can buy litmus paper cheaply from some chemists.

(b) **Colour comparator**: cheap water-testing kits can be bought from special chemical suppliers. They usually include a number of **liquid indicators**. A few drops of one of these colour indicators are added to a small water sample, and the new colour of the solution is compared with a set of **standard colours** supplied with the testing kit.

(c) **pH meter**: such equipment provides the easiest way for determining the water pH, even in the field, but it is relatively expensive. The pH value is directly read from the meter after placing **the glass electrodes** in a water sample. Such electrodes are very fragile and should be well protected when being transported. They should be accurately calibrated in buffer solutions of known pH, at regular intervals.

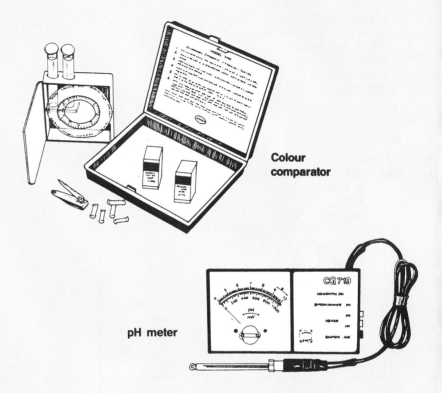

Colour comparator

pH meter

Note: because of the **variation of the pH** in fish ponds during the day (see below), you should **measure pH at a regular time**, preferably at sunrise. It is even better to measure the pH at regular intervals of two to three hours from sunrise to sunset, which will give you a good idea of the pH variations during daytime.

Choosing the pH value of your water

3. Fish production can be greatly affected by excessively low or high pH. **Extreme pH values** can even kill your fish. The growth of natural food organisms may also be greatly reduced. The critical pH values vary according to the fish species, the size of individual fish and other environmental conditions. For example, fish are more susceptible to extreme pH during their reproductive seasons, and eggs and young fish are more sensitive than adults.

4. Waters ranging in **pH from 6.5 to 8.5** (at sunrise) are generally the most suitable for pond fish production. Most cultured fish will die in waters with:

- pH below 4.5;
- pH equal to or greater than 11.

5. Fish reproduction can be greatly affected even at pH below 5.5, while a pH greater than 9 can already be detrimental to fish eggs and juveniles.

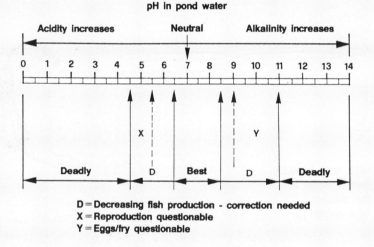

pH in pond water

Acidity increases | Neutral | Alkalinity increases

0 1 2 3 4 5 6 7 8 9 10 11 12 13 14

X | Y

Deadly | D | Best | D | Deadly

D = Decreasing fish production - correction needed
X = Reproduction questionable
Y = Eggs/fry questionable

Values of pH throughout the day and night

6. The original pH of the water may be affected by the pH of the soil (see Sections 41 and 42, **Soil**, **6**). However, the pH of pond water varies throughout the day mostly as a result of **photosynthesis**, and through the night through **respiration**.

(a) At sunrise, the pH is lowest.
(b) Photosynthesis increases as the light intensity increases. More and more carbon dioxide is removed from the water by the plants causing the pH to increase.
(c) A peak pH value is reached in late afternoon.

(d) Light intensity then starts decreasing, which reduces photosynthesis. Less and less carbon dioxide is removed from the water; as respiration adds more carbon dioxide to the water, pH starts to decrease.
(e) At sunset, photosynthesis stops, but respiration continues for the rest of the night. More and more carbon dioxide is produced, and pH keeps decreasing until sunrise, when it reaches its minimum.
(f) The next day, this **cyclic fluctuation** of pH starts again.

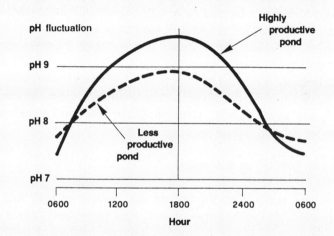

pH fluctuation

Highly productive pond

pH 9

pH 8

Less productive pond

pH 7

0600   1200   1800   2400   0600

Hour

7. This pH fluctuation varies in intensity. The more productive the pond, the richer its water will be in minute plant organisms (**phytoplankton**), the plant and animal respiration will be greater, and the daily fluctuation in pH will be stronger. The pH values of 9.5 will be quite common in late afternoon.

Correcting water of too low or too high pH

8. As you will learn later in this manual, pond water with a pH unfavourable for good fish production can be corrected.

(a) If the **pH is below 6.5 at sunrise**, you can use liming (see Chapter 5) and alkaline fertilizers (see Section 61).
(b) If the **pH is above 8.5 at sunrise**, you can use acid fertilizers (see Section 61).

11

23 Water turbidity and transparency

1. As you learned earlier, pond water contains **suspended particles** of different kinds. Water turbidity is caused by the presence of these suspended particles in varying quantities:

(a) **Mineral turbidity** is caused by a high content of silt and/or clay particles, which turn the water a light brown, sometimes reddish colour. It may occur when the water supply is turbid or a bottom feeding fish, such as the common carp, stirs up the bottom mud.

(b) **Plankton turbidity** is caused by a high content of minute plants and animals which colour the water in various shades of brown, green, blue-green or yellow-green, depending on which plankton species is dominant.

(c) **Humic turbidity** is caused by the presence of **humus** (see Section 16, **Soil, 6**), which turns the water a dark brownish colour. Its origin is usually the water supply, although it can also be caused by an excess of organic matter entering the pond.

The effects of turbidity in fish ponds

2. Mineral and humic turbidity reduce **the amount of light** that penetrates the water. In highly turbid waters, light penetrates only a short distance, and photosynthesis is reduced. Oxygen production during the daytime is relatively small. Both the growth of the fish and of their natural food organisms can be badly affected.

3. In addition, a **high mineral turbidity can affect fish directly** by injuring their breathing organs, reducing their growth rate or preventing their reproduction. In the same way, it can harm the minute animals called cladoceres and copepods (**zooplankton**), which are very important food for young fish (see Section 91, **Management, 21/2**).

With 10% turbidity light will reach the bottom of a pond

With 40% turbidity light will not reach the bottom of a pond

Measuring turbidity

4. Turbidity of pond water varies from almost zero to highly turbid, depending on the amount of suspended particles. The method used for its measurement varies according to the kind of turbidity present.

5. If it is **a mineral turbidity** (brownish water), you will need the help of a laboratory to determine the weight of material suspended in a given volume of water. This figure is called the **total suspended solids** (TSS), which is usually expressed in milligrams per litre (mg/l). When taking samples, be careful not to disturb the water too much, as you can increase the TSS very easily. Also, do not take the water only from the surface, as it is often much less turbid.

Amount of total suspended solids (TSS) present in pond water

TSS (mg/l)	Mineral turbidity
Less than 25	Low
25-100	Medium
Over 100	High

6. If it is **a plankton turbidity** (greenish water), you can estimate the level yourself using the two simple methods described below. They will also give you an estimate of the **potential fertility** of your pond, from which you can decide on the kind of management practice to be applied (see also Section 60).

Measuring plankton turbidity with your arm

7. This is a very simple method which does not require any special equipment. Proceed as follows.

(a) Slowly wade into the shallow part of your pond, trying not to disturb the pond bottom too much.

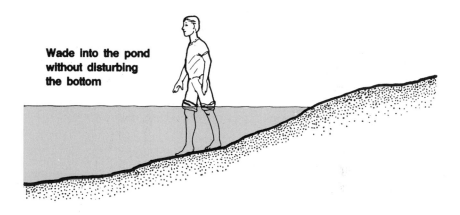

Wade into the pond without disturbing the bottom

(b) **Stretch one arm**, and immerse it vertically into the water until **your hand disappears from sight**.

(c) Note the water level along your arm:

- **if it is well below your elbow**, plankton turbidity is very high;
- **if it reaches to about your elbow**, plankton turbidity is high;
- **if it reaches well above your elbow**, plankton turbidity is low.

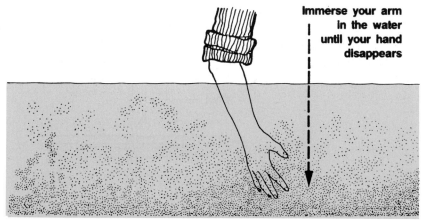

Immerse your arm in the water until your hand disappears

13

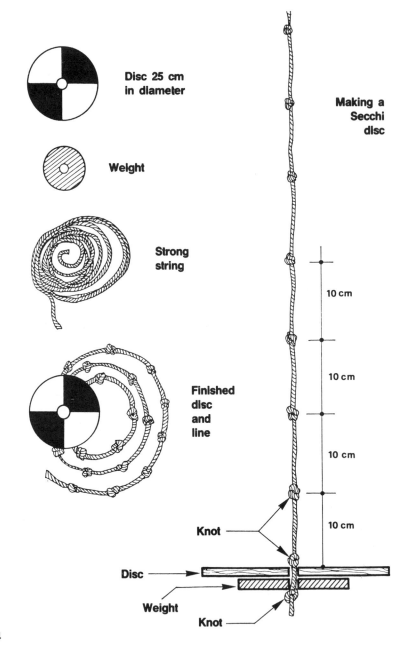

Disc 25 cm
in diameter

Weight

Strong
string

Finished
disc
and
line

Knot

Disc

Weight

Knot

Making a
Secchi
disc

10 cm

10 cm

10 cm

10 cm

Measuring turbidity with the Secchi disc

8. **The Secchi disc** is a very simple tool which can be used to give a better estimate of turbidity. It is particularly useful in green-coloured ponds to estimate plankton turbidity. This measurement is then called the **Secchi disc transparency**.

9. You can easily build a Secchi disc yourself. Proceed as follows.

(a) Cut **a round disc about 25 cm in diameter** from a piece of wood or metal, such as a pounded tin can for example.
(b) On its surface, mark **two lines** to make four quarters. Paint these black and white using matt paint to prevent glare.
(c) Drill a small hole at the centre of the disc. Through this hole **pass a line** or a piece of string about 1 to 1.5 m long.
(d) Below the disc, attach to the line a small **weight** such as a long bolt or a stone.
(e) **Fix the disc** at the bottom of the line, against the bottom weight, by knotting the line around a small piece of wood or metal, across the top of the disc.
(f) **Mark** the rest of the line with knots or tightly tied coloured thread at 10-cm intervals.

Note: instead of using a line, you may also attach the disc from its centre to a graduated vertical stick about 100 cm long.

Measuring the Secchi disc transparency

10. To measure the Secchi disc transparency, proceed as follows.

(a) Slowly lower the disc into the water.
(b) Stop when it just disappears from sight.
(c) Note at which point the line breaks the water surface. Mark this point A.
(d) After noting at which point along the line the disc just disappears, lower the disc a little and then raise it until it just reappears. Mark this point B.
(e) Mark point C, midway between points A and B.
(f) Measure the transparency of the water as equal to the distance from the top of the disc to this point C, counting the knots along the line. This figure is the Secchi disc transparency.

14

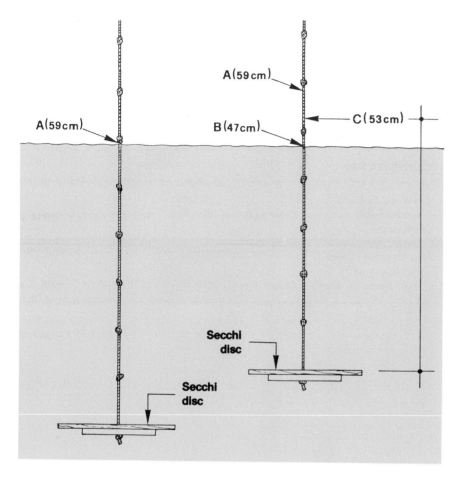

A(59cm)

A(59cm) B(47cm) C(53cm)

Secchi disc

Secchi disc

If **the Secchi disc transparency** is:

- **less than 40 cm**, there is too much plankton and your fish are in danger during the night when oxygen is not produced by photosynthesis and when too much oxygen is consumed by the respiration of this plankton;
- **40 to 60 cm**, the fish production will be the best;
- **over 60 cm**, there is too little plankton, and your fish do not have enough natural food to eat.

Controlling turbidity

12. There are several ways to control water turbidity, at least partly, depending on the kind of turbidity present.

(a) **To control mineral turbidity**, you may use:

- **a settling basin** (see Section 116, **Pond construction for freshwater fish culture**, *FAO Training Series*, **20/2**);
- **a water filter** (see Section 29);
- **organic matter** spread throughout the pond at the rate of 20 kg/100 m² (two to three treatments may be necessary);
- **alum** (aluminium sulphate) or **gypsum** (magnesium sulphate), at a rate of 1 to 3 kg/100 m², testing a small area first.

(b) **To control plankton turbidity**, you may use:

- **a water filter** (see Section 29);
- adequate **liming** (see Chapter 5);
- adequate **fertilization** (see Chapter 6).

11. **To obtain the best measurement**, take note of the following points:

(a) Measure transparency between 09.00 hours and 15.00 hours on calm days.
(b) Whenever possible, make the readings when the sun is out, not behind a cloud.
(c) Look at the sinking disc from directly above, if possible with the sun behind you.
(d) Keep the disc clean, particularly the two white quadrants. If necessary, repaint the disc black and white.

TABLE 1

Thermal ranges of common fish species (in °C)

Fish species	Dangerous pond-water temperature		Optimum thermal range for adults (opt. temper.)	Thermal range for spawning
	Lower	Upper		
WARMWATER FISH				
Micropterus salmoides Largemouth bass	2	35	23-30	17-20
Ictalurus punctatus Channel catfish	5	35	25-30	16-28
Cyprinus carpio Common carp	2	36	23-26 (25)	Above 18
Ctenopharyngodon idella Grass carp	—	32	23-28	15-30
Hypophthalmichthys molitrix Silver carp	—	32	23-28	15-30
Aristichthys nobilis Bighead carp	5	37	23-31	17-30
Carassius auratus Goldfish	5	37	25-30 (25)	Around 25
Clarias gariepinus African catfish	—	—	25-27	20-30
Tilapia aurea Blue tilapia	9	38	27-30	20-30
Tilapia nilotica Nile tilapia	12	38	27-30	22-32
Clarias batrachus Walking catfish (Asian)	15	—	29-32	22-32
Catla catla Catla (Indian carp)	15	34	26-29	22-28
Cirrhinus mrigala Mrigal	12	38	22-32	24-31
Labeo rohita Rohu	3	36	(28)	24-31

Fish species	Dangerous pond-water temperature		Optimum thermal range for adults (opt. temper.)	Thermal range for spawning
	Lower	Upper		
COLDWATER FISH				
Salvelinus fontinalis Brook trout	Close to 0	18	10-14 (13)	0-14
Salmo trutta Brown trout	Close to 0	20	12-15 (14)	0-15
Oncorhynchus mykiss Rainbow trout (syn. *Salmo gairdneri*)	Close to 0	22	15-17 (16)	4-18

The importance of water temperature for fish farming

1. The growth and activity of the fish depend on its body temperature. **The body temperature of fish** is about the same as the water temperature and varies with it. A relatively low water temperature can adversely affect fish by:

- slowing down the development of their eggs;
- reducing the growth of juveniles and older fish;
- delaying and even preventing their maturation and spawning;
- decreasing their food intake and even stopping it completely;
- increasing their susceptibility to infections and diseases.

2. Each fish species is adapted **to grow and reproduce** within well-defined **ranges of water temperatures**, but optimum growth and reproduction take place within narrower ranges of temperature. It is important, therefore, to know the water temperatures available at your fish farm well in order to select the right species of fish and to plan its management accordingly.

3. There are two main groups of fish (see **Table 1**):

- **coldwater fish**, those needing water temperatures below 15°C to breed, grow best at temperatures below 18°C and rarely survive long at temperatures above 25°C;
- **warmwater fish**, which need temperatures above 15°C to breed, grow best at temperatures over 20°C and can survive very high temperatures above 30 to 35°C.

4. Because the fish require sufficient dissolved oxygen in the pond water, the water temperature also affects **the breathing**, or respiration, of the fish. As you will learn in the next section, **the maximum quantity of dissolved oxygen** present in water depends on its temperature: the warmer the water, the less dissolved oxygen it can contain. For this reason, if the pond becomes too warm, the fish can run out of oxygen.

5. Fish have adapted to this situation by living in waters that provide them with sufficient oxygen, and so:

- trout which require a lot of oxygen live better in cold waters; and
- common carp and tilapias, which require less oxygen and cannot tolerate cold water, live better in warm water.

6. Water temperature also affects **other aquatic organisms** in the pond such as plankton, plants and animals. The pond fish may depend on them for food or for the production of oxygen through photosynthesis.

7. Temperature also affects the **density*** of water. The **density** of fresh water is at its **maximum at 4°C and decreases at higher or lower temperatures**, as shown in the graph below. This variation has several important consequences for fish ponds (see top of page 18).

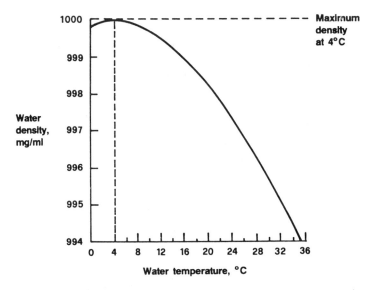

Effect of temperature on pure water density

(a) Water becomes lighter as it cools down **below 4°C**, therefore ice which forms at 0°C will float on the surface of the pond, and water below it will be warmer.

(b) Water also becomes lighter as it warms up **above 4°C**, therefore the warmest water is always at the top of a pond and the coolest water at the bottom.

(c) Over longer periods of warm weather, the warmer and lighter surface waters tend to form a separate layer from the colder and heavier bottom waters: **the pond water stratifies into distinct layers**.

(d) **In deeper ponds**, such as barrage ponds, such stratification may establish itself for a long period.

8. The pond water then forms three different layers:

- the upper, warmer and lighter **epilimnion**, in which the water temperature is relatively similar across the layer; the water is well mixed by wind, and usually has active photosynthesis and good oxygen levels;

- **the thermocline**, in which the water temperature drops and the density increases rapidly, thereby forming a sort of barrier which separates the pond water into two distinct parts;

- the lower, cooler and denser **hypolimnion**, in which the water temperature is relatively similar across the layer. The water cannot be mixed by wind any more, and in the absence of light and photosynthesis, dissolved oxygen gradually decreases, being mostly used for decomposition (see Section 20). It may even disappear completely from the bottom water, making life for fish and many other plants and animals impossible in this part of the pond. Because this area is separated from the surface waters, fertilizers or feed materials falling to the pond floor are no longer available for the plankton or the fish.

9. In cooler weather, **heavy cool rains or strong winds** can cause this water stratification to break up. The whole water mass then **turns over**, bringing the cooler, oxygen-poor bottom waters to the surface of the pond and sometimes killing the fish. In some cases, the nutrients and feed materials brought up from the bottom water can also cause excessive plankton growth.

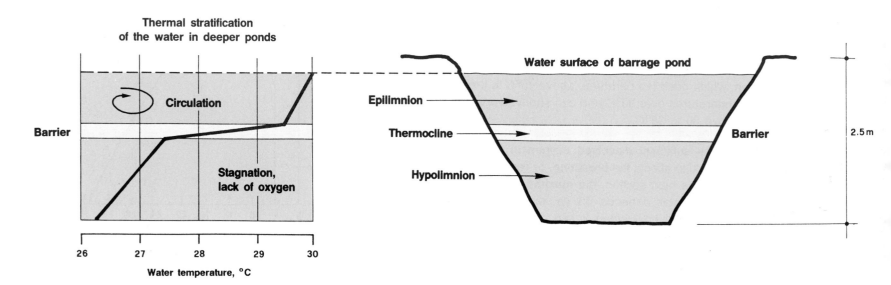

Thermal stratification of the water in deeper ponds

Circulation

Barrier

Stagnation, lack of oxygen

Water temperature, °C

Water surface of barrage pond

Epilimnion

Thermocline

Hypolimnion

Barrier

2.5 m

Measuring water temperature

10. To measure water temperature, you need a simple **thermometer**, graduated in degrees Celsius (°C) or centigrades. Such a thermometer is very fragile. To carry it around, you should keep it in a protective case. You can easily make one from a piece of bamboo in which you place some cotton wool.

11. To measure **the temperature of surface waters**, for example in the feeding canal at the pond water inlet or in the pond itself next to the outlet structure, proceed as follows.

(a) Place the bulb of the thermometer under water, at a depth of about 15 to 20 cm.
(b) Wait a short time until the column comes to a standstill.
(c) Without lifting the thermometer out of the water, read the temperature.
(d) Write this temperature down in a notebook for your records (see Chapter 16, **Management**, 21/2).

Note: you can use the same method with a bucket full of water, but be careful to measure the temperature immediately after the water is collected.

12. If you have to measure water temperatures often, it may be best to tie a sampling bottle to a pole with the bulb of the thermometer permanently attached just inside the bottle as shown in the illustration. Read the temperature as soon as you have filled the sampler with surface water.

13. To measure **the temperature of water from a greater depth**, for example bottom water in a pond near the outlet structure, you need **a better water sampler** such as the one described earlier (see Section 21). Then proceed as follows.

(a) Insert the thermometer into the sampler.
(b) After a few seconds, read the water temperature.
(c) Write it down in a notebook for your records.

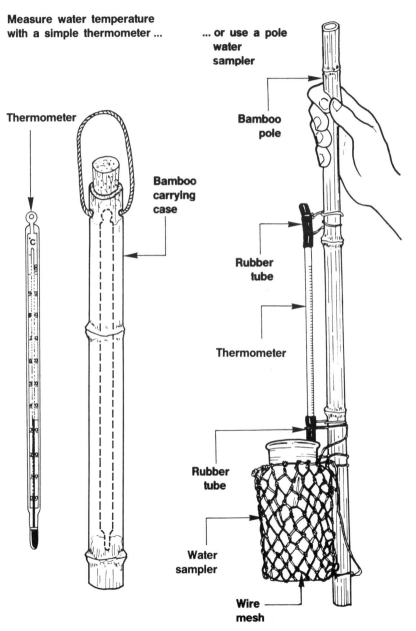

Measure water temperature with a simple thermometer ...

... or use a pole water sampler

Thermometer

Bamboo pole

Bamboo carrying case

Rubber tube

Thermometer

Rubber tube

Water sampler

Wire mesh

14. Whenever you need **to closely follow temperature changes in your pond for management purposes**, such as fish propagation (see Section 80, **Management, 21/2**), supplementary feeding (see Section 103, **Management, 21/2**) or protection of fish stocks against cold, it is best to measure the water temperature **twice a day**. The best time to do this is **shortly after sunrise**, when the air temperature is near its **minimum value**, and again **shortly after midday**, when the air temperature is near its **maximum value**, see graph below.

15. You can calculate **the average daily temperature** of the water and list thermal fluctuations in **a table** as shown in paragraphs 20 to 22 on the next page.

Surface water temperatures recorded from Pond 8 (°C)

Date	06.30 hrs	13.00 hrs
11.02.88	25.2	27.0
12.02.88	24.8	26.4
13.02.88	24.3	25.8
14.02.88	23.6	25.1
15.02.88	22.4	24.1
. . .	. . .	. . .

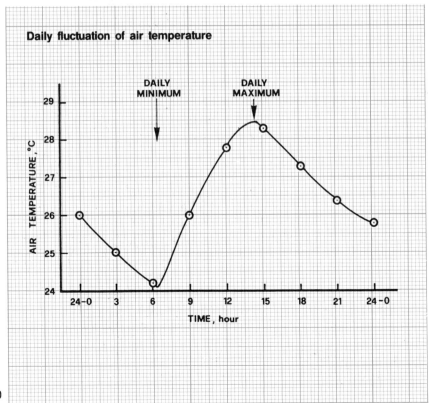

Minimum-maximum thermometer

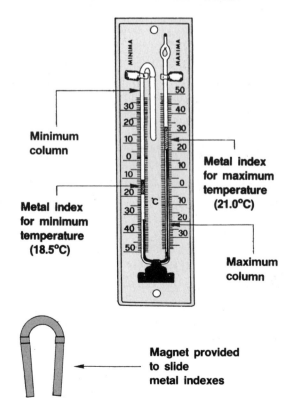

Recording extreme water temperatures

16. If you have **a maximum-minimum thermometer**, you will need to read the pond water temperature only **once a day** to obtain both the minimum and maximum temperatures.

17. First, **secure the thermometer in the pond**, within easy reach by:

- driving a wooden pole into the pond bottom, slightly angled from vertical, within easy reach of the outlet structure (the top of the pole should be above water);
- fixing a nail near the top of this pole, and attaching a string from the nail to the top of the thermometer;
- hanging the thermometer about 50 cm below the water surface, at a safe distance from the wooden pole.

18. To record **the extreme daily water temperatures**, proceed as follows.

a) Each morning, some time between 08.00 and 10.00 hours, pull the string up and lift the thermometer out of the water.

b) **At the bottom of the minimum metal index**, read the minimum temperature of the water, recorded since before sunrise that same morning. Note this down as **the minimum temperature for that day**.

c) **At the bottom of the maximum metal index**, read the maximum temperature of the water, recorded since midday of the previous day. Note this down as **the maximum temperature for the previous day**.

d) Using **the magnet** provided with the maximum-minimum thermometer, slide the two metal indexes down until they touch each of the mercury columns.

e) Hang the thermometer back in the water.

Note: you can also measure the temperature each evening. In this case, the minimum and maximum temperatures will both be for that day. If you have several ponds of about the same size and depth, fed with a common water supply, it is sufficient to obtain the minimum and maximum temperatures **in one of these ponds only**.

19. You can then calculate the **average daily temperature** of the water and show thermal fluctuations on **a graph**, as explained below.

Note: in a shallow pond not much more than 1 m deep, it is sufficient to measure the water surface temperature. In a pond deeper than 1.5 m, it is best to also measure the water temperature near the pond bottom.

Calculating the average daily temperature of pond water

20. This is a very simple calculation:

(a) For each day, add the minimum temperature to the maximum temperature.
(b) Divide this result by two to obtain **the average temperature** of each particular day.

Average surface water temperatures recorded from Pond 8 (°C)
(see also previous example on page 20)

Date	Minimum	Maximum	Daily average[1]
10.02.88	—	27.5	—
11.02.88	25.2	27.0	26.1
12.02.88	24.8	26.4	25.6
13.02.88	24.3	25.8	25.1
14.02.88	23.6	25.1	24.4
15.02.88	22.4	24.1	23.3
. . .	. . .	. . .	. . .

[1] The daily average is calculated as the sum of the minimum and maximum temperatures divided by 2

21. This method gives you a good estimate of the temperatures in which your fish live.

Note: if, for a deeper pond, you have also measured one water temperature near the bottom, you should calculate two average temperatures:

- **the surface water average**, as explained above; and
- **the bottom water average**, from the minimum and maximum temperatures observed near the bottom.

22. **The overall average of the pond water** is then estimated by dividing by two the sum of the surface and bottom water average temperatures.

Showing water temperature fluctuations in a graph

23. To guide your management practices, you can show on a graph **the variations of daily water temperature** over a period of time. Proceed as follows.

(a) Obtain some **square millimetric paper** or photocopy the sheet at the end of this manual.
(b) On **the horizontal axis**, show the time scale, adjusted according to the length of the period. Mark it with exact dates to avoid any error.
(c) On **the vertical axis**, show the water temperature scale, adjusted according to the expected range of temperatures and mark it with the temperature values (°C).
(d) Enter regularly on the graph the temperature values recorded in your notebook. For example, you may be interested in showing the fluctuations of:

   ● **the minimum water temperatures**, at the bottom of the graph;
   ● **the maximum water temperatures**, at the top of the graph;
   ● **the average water temperatures**, in the middle of the graph.

24. When this information is plotted on a graph, you can easily follow the changes in water temperature that are affecting your fish stocks.

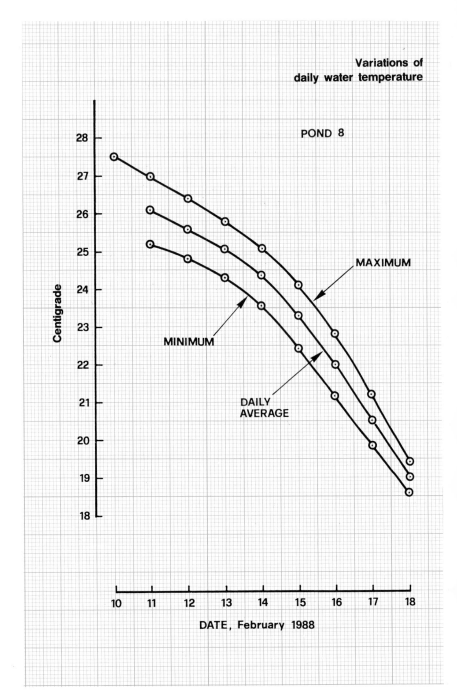

Variations of daily water temperature

POND 8

MAXIMUM

MINIMUM

DAILY AVERAGE

Centigrade

DATE, February 1988

25. It is possible to take some action to improve, within certain limits, the water temperature in fish ponds. But remember that:

- **the larger the pond**, the more stable its average water temperature will be, whereas in very small ponds of only a few hundred square metres, temperature conditions can change very rapidly, for example during a rainstorm or because of a strong, cold wind;
- **the smaller the pond**, the easier it will be to change its average water temperature through management.

26. If you wish **to increase the average water temperature** of your pond, for example for early spawning of warmwater fish or for prolonged growth or survival over winter months, you can do so in the following way.

(a) Plant wind-breaks across the direction of seasonal cooling winds (see Section 42).

(b) To take advantage of sunny weather, build shallower ponds which warm up faster.

(c) If you have a cold water supply, warm it up using a shallow warming pond placed before the main pond.

(d) For overwintering during a cold season, build deeper ponds, which are less affected by sudden weather changes. If ice forms on the pond surface, bottom waters will remain warmer, at about 4°C when density is the greatest (see Section 24).

(e) If a warmer supply of water is available, **discharge cooler bottom water** using the monk outlet in the way described in a previous manual (see Section 107, **Pond construction, 20/2**).

27. If you wish **to decrease the average water temperature** of your pond, for example to increase the overall dissolved oxygen content or to avoid the effects of high temperatures, it is better to increase the inflow of cooler water while:

- **in shallow ponds**, discharging the warmest surface water;
- **in deeper, stratified ponds**, discharging the most deoxygenated bottom waters from the hypolimnion. Be careful, however, not to mix the layers while doing so.

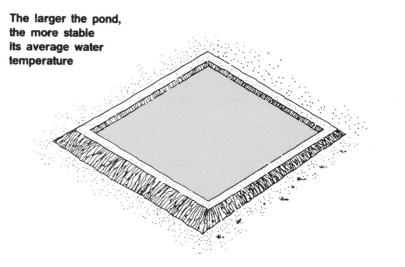

The larger the pond, the more stable its average water temperature

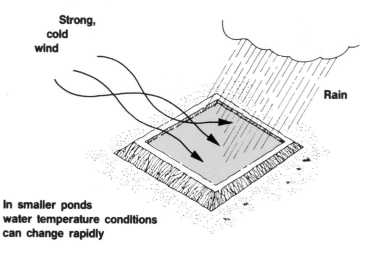

Strong, cold wind

Rain

In smaller ponds water temperature conditions can change rapidly

25 Dissolved oxygen in fish ponds

1. The most important gas dissolved in water is oxygen (O_2). As you have already learned, **dissolved oxygen (DO)** is essential to most living organisms for their **respiration**. Oxygen is also necessary for dead organic matter to be broken down during the process called **decomposition**.

Sources of dissolved oxygen

2. The oxygen dissolved in water has two sources:

- atmospheric oxygen;
- photosynthesis.

3. **The atmospheric oxygen** in contact with the water surface is an unlimited source of oxygen; unfortunately, its passage into water, its **diffusion** and its subsequent **dissolving** into water is a very slow process. You will learn how to improve this process by using **aerators** later in this section.

4. The major source of dissolved oxygen in ponds is **photosynthesis** (see Section 20). Remember that this process depends on **the amount of light** available to the plants and therefore:

- oxygen production decreases during cloudy days;
- it completely stops at night;
- it gradually decreases as water depth increases and light levels diminish, the rate of the decrease depending on the water turbidity (see Section 23).

Measuring DO content in water

5. You can measure how much oxygen is dissolved in water either by **chemical methods** or by **electrical methods**.

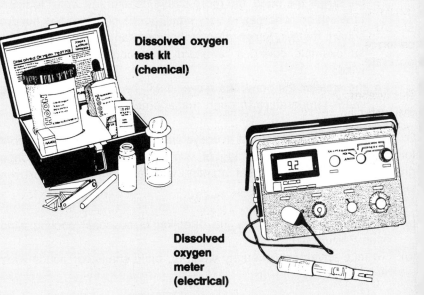

Dissolved oxygen test kit (chemical)

Dissolved oxygen meter (electrical)

6. **Chemical methods** usually rely on the use of simple kits, which can be bought from specialized suppliers. They contain all the chemicals and small equipment necessary to determine the DO content with sufficient accuracy for pond management purposes. Instructions should be closely followed. The measurement is taken on a small water sample obtained with the water sampler described above.

7. **Electrical methods** are based on the use of **an oxygen meter**, a rather expensive device which can be bought from specialized suppliers. It has the advantage that the DO content can be measured directly from the water, at any depth. The DO content is read from a scale. Instructions should be accurately followed. You should take particular care to calibrate (i.e. adjust) the meter regularly to make sure it remains accurate.

Remember: when measuring DO content, you should always measure **the water temperature simultaneously**, so that you will be able to relate the DO measurement to this temperature.

8. **When to measure DO** depends on the purpose of the measurements.

(a) If you plan to measure DO **routinely** as part of the regular monitoring of the fish farm (see Chapter 16, **Management, 21/2**), it is best to measure it twice on a specific day:

- just after sunrise, when DO is at its minimum;
- a few hours later.

(b) In seasons when DO might be insufficient, if you want **to predict low night-time DO content**, you should also measure it twice:

- just before sundown;
- a few hours later.

Apply the method described below.

(c) **If you suspect from observation that there might be a lack of DO**, you could measure the DO content immediately to confirm and then take remedial action.

Determining the average DO content of the water in a pond

9. To estimate **the average DO content** of the water in a shallow pond at a certain time, you should obtain a series of water samples.

10. The simplest (but less accurate) way is to obtain water samples from **one station only** but at **different depths**, as follows.

(a) Select the sampling station at the deep end of the pond and a little away from the dikes, for example in front of the monk outlet structure.
(b) Take a first sample from about 30 cm below the water surface and measure its DO content = A.
(c) Take a second water sample from a depth = 0.50 × total depth and measure its DO content = B.
(d) Take another water sample from a depth = 0.80 × total depth and measure its DO content = C.
(e) Calculate the **average DO content** of the water in the pond as: $X = (A + B + C) \div 3$.

11. For more accurate results, proceed as just described, but use **two stations**:

- one in **the middle** of the pond, to obtain a first average DO value = X_1;
- one at **the deep end** of the pond, to obtain a second average DO value = X_2;
- calculate **the overall average DO content** of the water in the pond as $X = (X_1 + X_2) \div 2$.

For more accurate results measure DO content using two stations

25

12. Do not take water samples near growing aquatic plants or beneath a heavy cover of algae blown against the shore, as these samples would not be typical of the conditions in the rest of the pond.

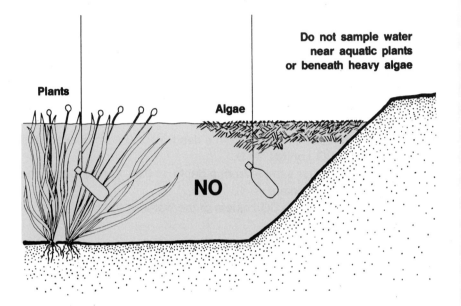

Do not sample water near aquatic plants or beneath heavy algae

Plants

Algae

NO

13. If your pond is large and deep, such as **a barrage pond**, you may need more water samples to obtain a better estimate of the average DO content of water in the pond. You should, if needed:

- add one sampling station at the shallow end of the pond;
- obtain at each station four samples instead of three, at depths equal to 0.1, 0.25, 0.50 and 0.75 times the total depth at the station; and
- calculate the averages accordingly.

14. In some cases (see below), you may wish to obtain the DO content of the upper layers of the pond only. For this you should take samples in this upper layer only, and calculate the averages as before.

Expressing the DO content of the water

15. The concentration of DO in water may be expressed in several ways:

(a) As **the weight of oxygen** per volume of water, for example:

- in milligrams per litre (mg/l);
- in grams per cubic metre (g/m^3);
- in parts per million (ppm), where **1 ppm = about 1 mg/l**.

(b) As **the volume of oxygen** per volume of water, most generally in millilitres per litre (ml/l), where **1 ml/l = 0.7 mg/l**.

(c) As **the oxygen saturation value**, the percentage of the maximum amount of oxygen that the water can hold at that particular temperature, see paragraph 19 on next page.

Example

A water sample at 30°C has a DO content = 6 mg/l. You can express this DO content in other units such as:

- 6 mg/l is roughly equal to 6 ppm;
- 6 mg/l ÷ 0.7 = 8.6 ml/l;
- at 30°C, the 100-percent saturation is 7.54 mg/l. Then **the oxygen saturation value** of this sample is equal to (6 mg/l ÷ 7.54 mg/l) × 100 = 79.6 percent.

16. For semi-intensive fish farming, the DO content of water is generally expressed either in mg/l or in percent saturation.

Determining how much oxygen water can hold

17. As for all other gases, the maximum quantity of oxygen that water can hold, **the oxygen solubility in water**, depends on three factors.

(a) **Temperature**: the warmer the water, the less oxygen it can hold.

(b) **Atmospheric pressure**: the lower the pressure, the less oxygen can be held and therefore:

- the higher the altitude, the less oxygen water can hold;
- during low atmospheric pressure conditions, e.g. stormy weather, oxygen solubility decreases;
- oxygen solubility increases with water depth.

(c) **Salinity**: the more saline the water, the less oxygen it can hold.

18. The maximum quantity of oxygen which a particular body of water can normally hold is called **the 100-percent saturation value**. Under certain circumstances, it may happen that the pond water contains more than this 100-percent saturation value. It is then said that there is **supersaturation** of the water with oxygen, a phenomenon which may happen, for example, in the afternoon when photosynthesis has been very active.

19. Some **100-percent saturation values** (mg/l) for oxygen in water for various temperatures, altitudes, water depths and salinities are given in the following charts to show you the expected variations of the DO content under various circumstances.

Variation of oxygen saturation with water temperature

Water temperature (°C)	DO 100% sat.[1] (mg/l)	Water temperature (°C)	DO 100% sat.[1] (mg/l)
0	14.60	18	9.45
2	13.81	20	9.08
4	13.09	22	8.73
6	12.44	24	8.40
8	11.83	26	8.09
10	11.28	28	7.81
12	10.77	30	7.54
14	10.29	32	7.29
16	9.86	34	7.05

[1] For fresh water, at sea level

Variation of oxygen saturation with altitude and water depth[1]

Altitude (m)	Water temperature 20°C	Water temperature 30°C	Water depth (m)	Water temperature 20°C (at sea level)
0	9.08	7.54	0	9.08
300	8.76	7.27	0.5	9.53
600	8.46	7.01	1.0	9.98
900	8.16	6.77	1.5	10.43
1 200	7.88	6.53	2.0	10.87
1 500	7.61	6.29	2.5	11.32
1 800	7.34	6.07	3.0	11.77
2 100	7.08	5.85	3.5	12.22

[1] 100-percent saturation values for fresh water, in mg/l

Variation of oxygen saturation with water salinity (mg/l)[1]

Water salinity (ppm)	Water temperature 20°C	Water temperature 30°C
0	9.08	7.54
5	8.81	7.33
10	8.56	7.14
15	8.31	6.94
20	8.07	6.75

[1] 100-percent saturation values, at sea level

Example

If the DO content of surface water is measured equal to 8.2 mg/l in a freshwater pond located at 300 m altitude and if the water temperature is 30°C, you find from the chart (Variation of oxygen saturation with altitude and water depth) in the centre of the next column that the 100-percent saturation value is 7.27 mg/l. **The oxygen saturation** value of the water is therefore equal to (8.2 mg/l ÷ 7.27 mg/l) × 100 = 112.8 percent, which means that this water is **supersaturated** in oxygen.

Using a graph to determine oxygen saturation values

20. It is often easier to use the following **graphical method**, which gives you a quick estimate of the oxygen saturation value good enough for freshwater pond management purposes. Proceed as follows.

(a) Measure the DO content of the water, in mg/l.
(b) Correct this value for **altitude** by multiplying it by the appropriate **correction factor** to obtain the DO content at sea level, as follows.

Altitude (m)	Correction factor		Altitude (m)	Correction factor
0	1.00		1 000	1.12
100	1.01		1 100	1.14
200	1.02		1 200	1.15
300	1.04		1 300	1.17
400	1.05		1 400	1.18
500	1.06		1 500	1.19
600	1.07		1 600	1.21
700	1.09		1 700	1.22
800	1.10		1 800	1.24
900	1.11		1 900	1.25

(c) Introduce this sea-level DO value on **the bottom horizontal line** using the **graph** shown on page 29 to obtain **point A**.
(d) On **the top horizontal line** of the nomograph, determine **point B** at the water temperature (°C) value.
(e) Using a ruler, join pont A to point B to determine **point C** on the oblique central line.
(f) At **point C**, read the **DO saturation value** in percent.

Note: if you can, make a photocopy of the graph on page 29 so that you can use it again.

21. Remember: if the DO saturation value is higher than 100 percent, there is **oxygen supersaturation** in the water.

Example

You have measured a DO content = 5.4 mg/l in a shallow pond located at an altitude of 275 m above sea level, and the water temperature was 28.3°C. Determine the DO saturation value as follows:

- for altitude = 275 m, the correction factor is about 1.03;
- the DO content at sea level would then be 5.4 mg/l × 1.03 = 5.56 mg/l;
- determine point A = 5.6 mg/l on line A;
- determine point B = 28.3°C on line B;
- with a ruler, determine point C = 71 percent.

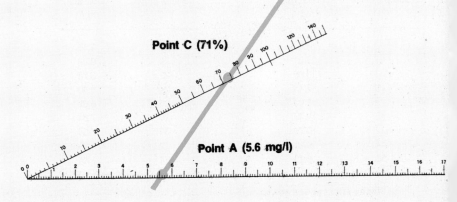

GRAPH 1

Graph for the determination of oxygen saturation values in fresh water at sea level

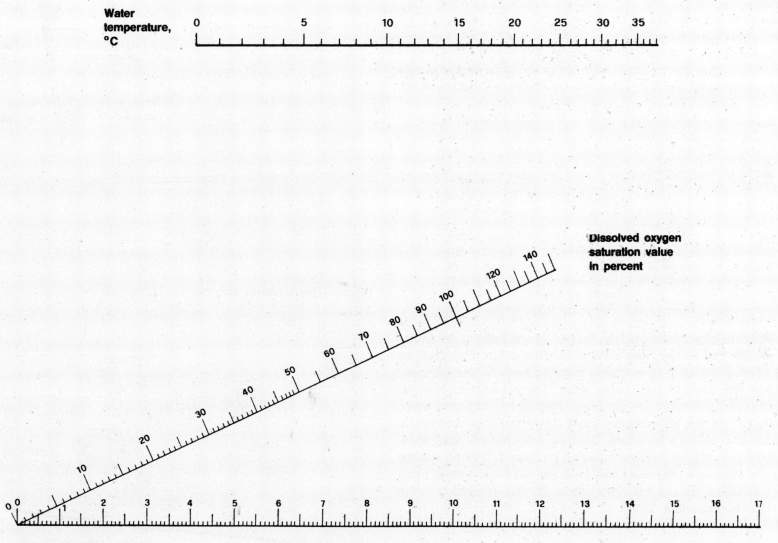

Water temperature, °C

Dissolved oxygen saturation value in percent

Dissolved oxygen, mg/l

22. The oxygen requirements of fish are determined by three **basic factors**:

- fish species;
- size of the fish; and
- water temperature.

23. In addition, there may be some variation due to **physiological factors** such as activity, feeding and digestion, sexual maturation and spawning.

DO requirements at various stages of life (in mg/l or percent saturation values)

Fish species	Eggs and juveniles	Adults	
		Minimum DO level	Preferred DO level at least equal to
Trout	Close to 100%	5 mg/l (50%)	8 mg/l or 70%
Common carp	At least 70%	3 mg/l (30%)	5 mg/l or 50%
Tilapia	At least 70%	2 mg/l	4 mg/l or 50%
African catfish	At least 90%	1 mg/l or less (aerial respiration)	3 mg/l or 35%

24. As you have already learned (see Section 24 and **Table 1**), coldwater fish require higher oxygen levels than warmwater fish. Fish such as catfish, which are used to slow-moving water bodies, can tolerate lower levels than fish used to fast-moving water. For a particular species, younger fish require higher oxygen levels than adults. At higher water temperatures, the fish will consume more oxygen for their respiration. This factor can be very important, because when temperature rises, water holds less oxygen (see the top chart on page 27). When **actively feeding** and later, when **digesting their food**, fish will require much more oxygen than usual. You will learn more about DO requirements of fish in Section 140, **Management, 21/2**.

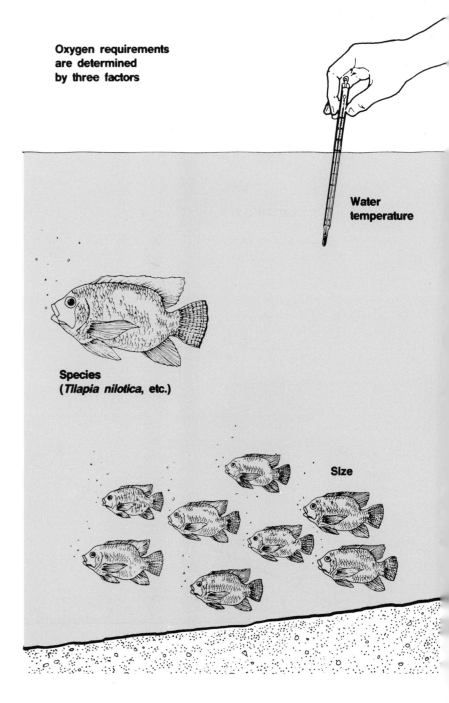

Oxygen requirements
are determined
by three factors

Water
temperature

Species
(*Tilapia nilotica*, etc.)

Size

Fluctuating oxygen levels

25. Two types of fluctuation in oxygen level can be found in fish ponds:

- **daily fluctuations**, both in surface water and in deeper water;
- **seasonal fluctuations**, mostly observed in deeper ponds.

26. **In surface water, the daily fluctuation** of the DO content is related to the 24-hour cycle of day and night.

(a) **From sunrise to sunset**, photosynthesis increases the DO level. On clear days, DO production is higher than on cloudy days. The higher the phytoplankton population, the higher the DO production.

(b) **At night**, photosynthesis does not take place, and therefore respiration reduces the DO content until sunrise. The higher the plankton population, the faster the DO will fall.

DO content in the surface water of a shallow pond during a 24-hour cycle (water temperatures from 28 to 33°C)

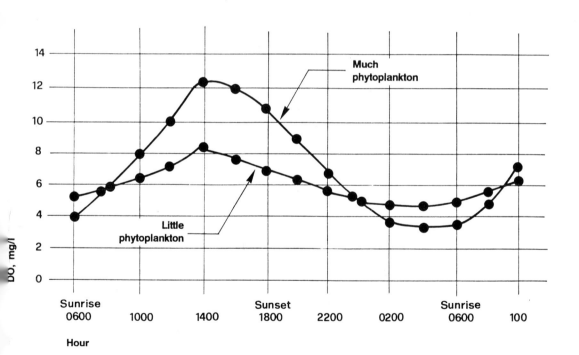

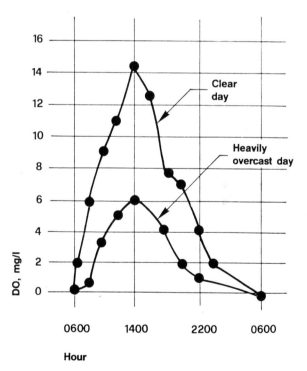

27. **In very rich ponds**, surface water may become **supersaturated** by midday. However as respiration is also high, there may be very little oxygen left by the end of the night. Your fish may die if you do not correct these conditions.

28. **In deeper water, the daily fluctuation** of the DO content is related to the plankton turbidity: the higher the turbidity, the smaller the amount of light penetrating deeper in the water, and the less the production of DO through photosynthesis in deeper water. **The DO content therefore decreases as depth increases**.

29. **In very rich ponds**, where there is a dense plankton population and high turbidity, the DO content of the lower depths of the water may become very low even during the day. The fish may have to concentrate at the surface of the pond to survive. Greater problems are to be expected after sunset.

30. **The seasonal fluctuation** of the DO content is essentially related to **the thermal stratification of the water** (see Section 24). As the thermocline establishes itself and restricts exchanges between the upper and lower layers, the DO content of the bottom water decreases, mainly because of the decomposing organic matter. It is only after the pond water has turned over that DO is brought back from the pond surface to the bottom through the general mixing of the water.

31. **In deep ponds rich in organic bottom mud**, the bottom water may become totally devoid of oxygen (anoxic) within a few weeks, and fish will not be able to live there. Later, when the pond water turns over, this anoxic water may reach the surface, together with decomposed organic material; many fish may die if you do not help them.

Daily fluctuations of DO content with water depth in a shallow pond with the presence of a high plankton turbidity (water temperatures from 26 to 33°C)

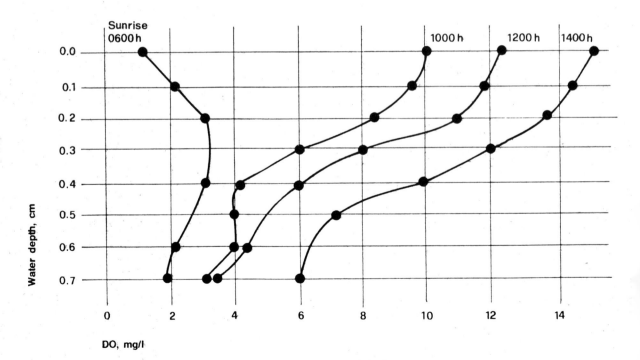

Sudden drops in the DO content of pond water

32. Apart from the fluctuations of the DO content described above which take place regularly, every day or seasonally, **the DO content of pond water may also decrease suddenly** for several other reasons. If this should happen, check on the following possible causes one by one:

(a) **The water supply**:

- **the incoming water** may have a very low DO content; water from a well or deep water from a reservoir may be low in DO; respiring or decomposing algae upstream in a surface water supply may reduce DO; DO may be lowered by organic pollution;
- **the water inflow** is too small.

(b) **The weather**:

- **the pond water** was stratified (see Section 24) and a **cool rain or a strong wind** has caused the pond water to turn over, bringing the deoxygenated bottom water to the surface;
- several days of **cloudy skies** or **rainy weather** have reduced oxygen production through photosynthesis;
- a period of **very hot days** has increased the water temperature, reducing DO saturation and increasing fish requirements of oxygen.

(c) **The fish pond**:

- the **plankton turbidity** is too high, and too much oxygen is consumed at night for respiration;
- there is too much **decaying organic matter**, and too much oxygen is used up for its decomposition.

(d) **Fish stock management**:

- there are **too many fish** in the pond;
- the fish are **overfed** resulting in fish wastes (faeces), and/or **unconsumed food** is decaying on the bottom.

33. Generally, **several factors** combine to cause a sudden drop in the DO content. In most cases, good management can prevent this.

The signs of low DO content in fish ponds

34. In the absence of a suitable chemical kit or oxygen meter, **you can observe signs** which tell you that there is not enough oxygen in your pond, for example:

- **tadpoles** assembling at the margin of the pond;
- **water snails** crawling up emerging plants;
- **an odour of rotten egg** rising from the water;
- the fish not **feeding well** or even stopping feeding;
- the fish coming **to the water surface** in an effort to breathe from the thin, better oxygenated surface film — this behaviour is called piping.

Predicting low DO content with a Secchi disc

35. In Section 23, you learned how to measure the Secchi disc transparency. If **phytoplankton** is the main source of turbidity in your pond, you can use this measurement together with other observations to predict low DO content. Proceed as follows.

(a) Measure Secchi disc transparency (SD).
(b) **If SD is smaller than 25 cm**, the risk of oxygen depletion is great, especially if:

- there is a heavy **cloud cover** for two to three consecutive days;
- there is a heavy **fog cover** in the morning.

(c) **If SD is between 25 cm and 60 cm**, there is still a risk, although decreasing as SD increases, that oxygen depletion may take place. Check on the weather and look regularly for signs at the pond.
(d) **If SD is greater than 60 cm**, the risk that oxygen depletion may suddenly take place is minimal, unless overcast weather persists for more than a week.

Note: when applying this method, you should remain aware of **the other reasons** that may cause a sudden DO drop.

36. If you have a kit or a meter to measure DO content, you may apply **the projection method** developed by Prof. C.E. Boyd and colleagues at Auburn University, USA (*Transactions of the American Fisheries Society*, 107: 484-92, 1978). It can be used in any pond but it is particularly useful when most of the water turbidity does not come from phytoplankton and when the Secchi disc method cannot be used reliably. Proceed as follows, referring to the graph on page 35:

(a) Measure the average DO content of the pond **at sunset**, for example at 18.00 hr, to determine X mg/l.

(b) Measure the DO content again **two to three hours later**, for example at 21.00 hr to determine Y mg/l. This value should be smaller than the first one.

(c) Using graph paper (examples of graph paper which you can photocopy are found at the end of this manual), **plot X and Y** values as DO contents (vertical scale) against time (horizontal scale).

(d) Join the two points **X and Y** by a straight line.

(e) **Continue this straight line down** to the bottom scale. This will **predict the average DO content** of the water in the pond later during the night.

37. You can now use this graph according to the kind of fish (species, size) in the pond and their minimum acceptable oxygen level (see the chart on page 30). **Draw a horizontal line** showing on the graph the minimum DO level acceptable in the pond. The following outcomes may thus be predicted.

(a) **If the projected line** (such as line C in the example) **intersects** this horizontal line after sunrise, there should be sufficient DO for the fish throughout the night.

(b) **If the projected line** (such as line B in the example) **intersects this horizontal line** just before or at **sunrise but also intersects the zero-DO line just after sunrise**, there will most probably be sufficient DO for the fish until sunrise. However, it might be advisable to improve the oxygenation of the water (described below), especially during the last part of the night.

(c) **If the projected line** (such as line A in the example) **intersects the horizontal line and the zero-DO line before sunrise**, there is a major risk that there will be a severe lack of DO during the night. Measures should be taken immediately to improve the water oxygenation.

Note: the results depend a lot on the accuracy of your calculations. To obtain good water samples and calculate the average DO content of the water in the pond, proceed as described in Sections 21 and 25.

38. During critical seasonal periods and in ponds where you expect problems, you can use **the chart method** over a longer time period. This is particularly convenient if you have an oxygen meter that gives you fast and reliable measurements of DO contents at any water depth. Proceed as follows, referring to graphs on pages 36 and 37:

(a) Measure the average DO content of the water in the pond at **sunrise**, for example at 06.00 hours, to determine X mg/l.

(b) Measure the average DO content of the water in the pond at **sundown**, for example at 18.00 hours, to determine Y mg/l.

(c) **Prepare a chart** with two scales showing time (horizontal) and DO content (vertical) **for each pond** you plan to monitor.

(d) Enter your measurements X and Y in the chart of the pond.

(e) Note also the weather conditions.

(f) **Repeat this procedure every day** during the period you are examining, or until oxygen problems cease.

39. By observing **the fluctuations of the average DO content** and the effect of daily weather conditions, you should be able to predict when it will become necessary to intervene to improve the oxygenation of the water and to avoid fish production losses. In particular, **you should observe**:

- how great **the daily fluctuations of DO content** are;
- how much DO content is present **at sunrise** and **at sundown**;
- if there is any gradual **downward trend** in DO content at sunrise and at sundown; and
- if particular weather conditions such as heavy rainfall or periods of cloud affect oxygen patterns.

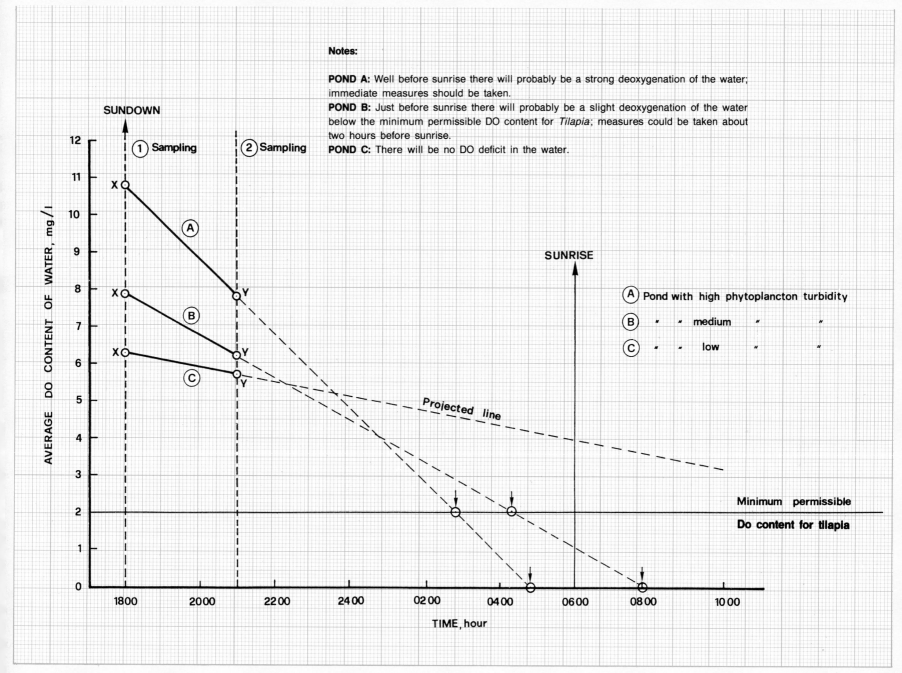

Notes:

POND A: Well before sunrise there will probably be a strong deoxygenation of the water; immediate measures should be taken.

POND B: Just before sunrise there will probably be a slight deoxygenation of the water below the minimum permissible DO content for *Tilapia*; measures could be taken about two hours before sunrise.

POND C: There will be no DO deficit in the water.

(A) Pond with high phytoplancton turbidity

(B) " " medium " "

(C) " " low " "

SUNDOWN

① Sampling   ② Sampling

SUNRISE

Projected line

Minimum permissible

Do content for tilapia

AVERAGE DO CONTENT OF WATER, mg/l

TIME, hour

Example

In pond 7, there is a downward trend in DO content both at sunrise and sundown; DO content at sundown is getting low; daily fluctuations are large: there is **too much phytoplankton present**, and a DO deficit may occur at night from 16 to 18 March on. Oxygenation will then have to be improved mechanically.

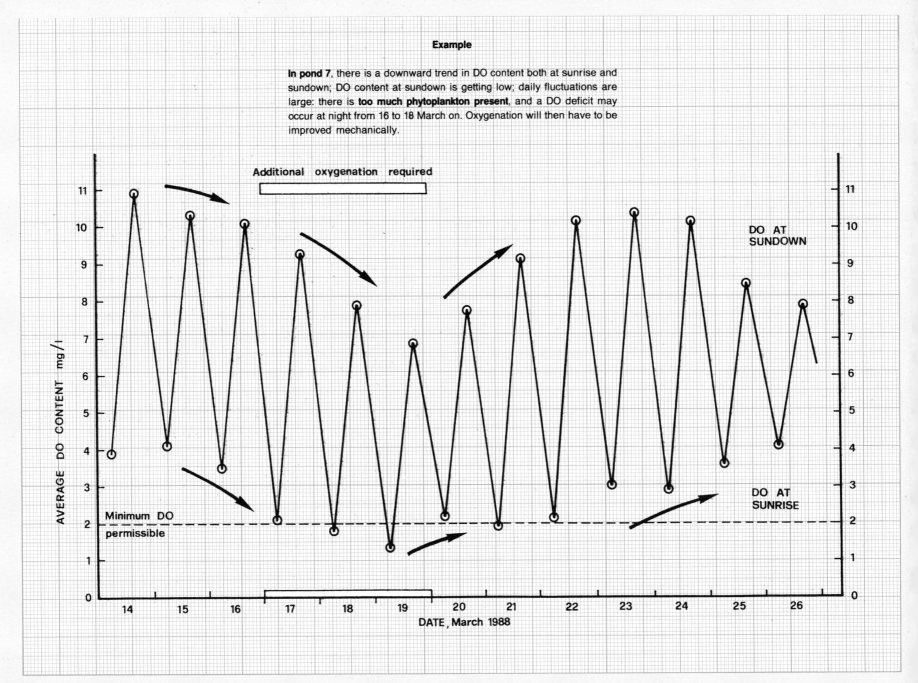

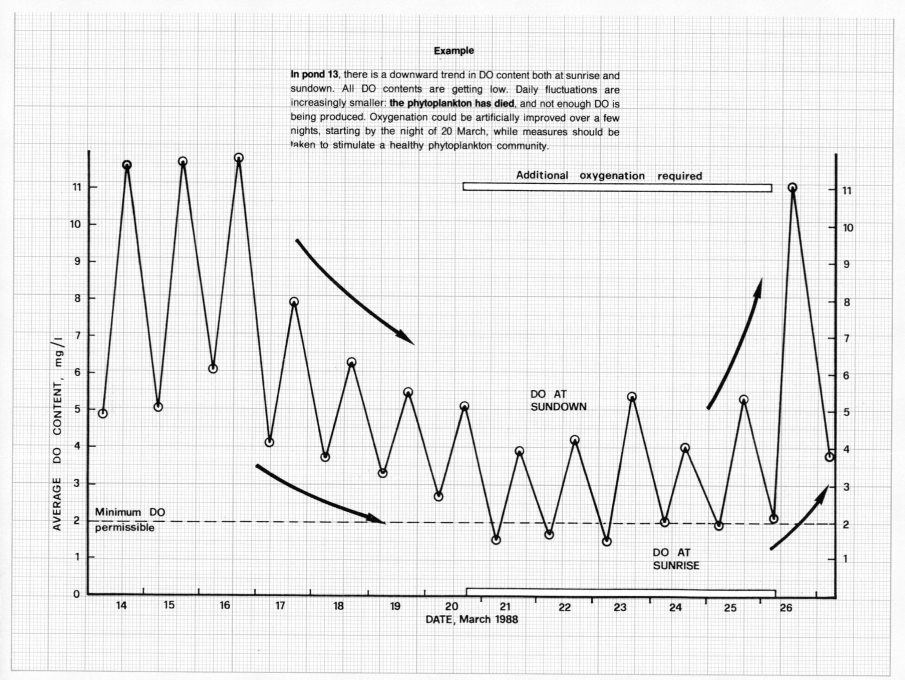

Example

In pond 13, there is a downward trend in DO content both at sunrise and sundown. All DO contents are getting low. Daily fluctuations are increasingly smaller: **the phytoplankton has died**, and not enough DO is being produced. Oxygenation could be artificially improved over a few nights, starting by the night of 20 March, while measures should be taken to stimulate a healthy phytoplankton community.

Additional oxygenation required

DO AT
SUNDOWN

AVERAGE DO CONTENT, mg/l

Minimum DO
permissible

DO AT
SUNRISE

DATE, March 1988

40. **The DO content of pond water can be increased** in several ways:

- directly **in the pond itself**, through design and management (this approach will be discussed in the next part of this section);
- **at the pond inlet**, through simple, cheap structures (see Section 26);
- **upstream from the pond inlet**, either before the main water intake or somewhere along the feeding canal, through cascades (see Section 27);
- if power is available, with use of **various mechanical devices** for the emergency aeration of pond water (see Section 28).

41. A simple way to ensure a good supply of atmospheric oxygen to fish ponds from the beginning is to **plan their design** to take maximum advantage of the **wind**. This will result in better aeration of the surface waters and better mixing with deeper water. As far as possible (see **Pond construction for freshwater fish culture**, *FAO Training Series*, **20/1**), you should design your ponds, particularly **the fattening ponds**, keeping the following points in mind.

(a) Find out which are the **most important winds** during the year, considering factors such as the critical times for pond oxygenation, the strength and continuity of the wind and the time of day the wind blows.

(b) **The longer the path of the wind** on the pond surface, the greater the wind action. If possible **orient your ponds** so that the longest dikes are parallel to the direction of the winds you want to use.

(c) It is easier for the wind to keep ponds well mixed when **they are shallow** rather than when they are deep.

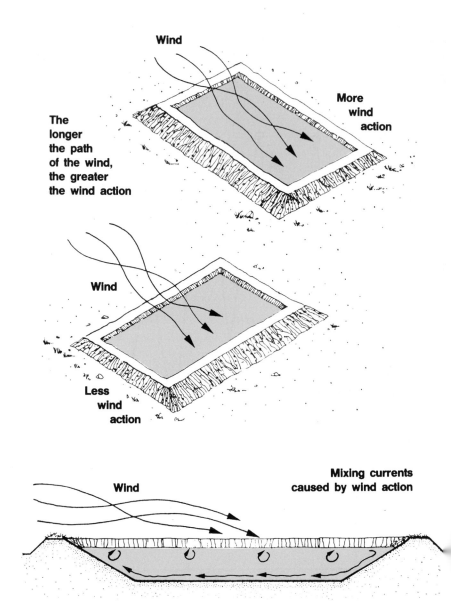

Wind

The longer the path of the wind, the greater the wind action

More wind action

Wind

Less wind action

Wind

Mixing currents caused by wind action

42. If you can identify the likely cause of the problem, **good pond management** may improve the DO content of the water. You should take the following routine measures, preferably **before any emergency** happens.

(a) **Direct improvement of DO content by**:

- increasing the inflow of well-oxygenated and/or cooler water;
- removing the less oxygenated bottom water, by using the pond outlet as explained earlier, and replacing it with better oxygenated water;
- in smaller ponds, increasing the mixing of air and water by splashing the water with your hands or with a broad stick or paddle.

b) **Reduction of oxygen consumption from decomposition of organic matter by**:

- avoiding overfeeding (see Section 96), if feeds are used;
- if possible, trying to drain away some of the loose organic matter on the bottom;
- if the pond is drainable, and you have drained the pond, keeping its **bottom thoroughly dry** for at least two to three weeks under tropical climates or three to four months under temperate climates to allow the humid, soft mud to become converted into normal soil and the **atmospheric oxygen** to promote processes such as the decomposition/mineralization of the organic matter and deacidification of the pond bottom;
- curing the drained pond by **removing some of the accumulated bottom mud** when it becomes too thick (you can profitably use such organic mud to fertilize land crops); and
- before refilling the pond with water, treating the bottom soil; **liming** (see Section 54) and **fertilizing** (see Chapter 6) **the pond bottom** can greatly improve its quality.

(a) To improve the effectiveness of drying the bottom of the pond, hoe or plough the soil to a depth of 15 to 20 cm.

(b) Vegetables or other crops may be cultivated together with pond drying to increase its benefits.

(c) Thorough drying is particularly advisable if the loose mud layer is thicker than 15 cm and if it smells putrid.

(d) Do not keep a fish pond thoroughly dry if:

- little or no mud has formed over pond bottoms which are too sandy or clayey, as the clay will crack and fissures will develop which could increase water seepage;
- little or no mud has been formed over an acidic or potentially acidic soil (see **Soil, 6**);
- a thick layer of non-putrid mud is present in a rich pond, which is well oxygenated from surface to bottom for most of the year.

(c) **Increased production of oxygen through photosynthesis by**:

- increasing **sunlight availability** by keeping shady trees and aquatic vegetation under control;
- improving **sunlight penetration** and reducing **night-time respiration** through the control of submerged plants and algae (see Section 49) whenever necessary;
- reducing the existing **phytoplankton** population slightly if it is causing oxygen levels to fluctuate too strongly (drain off some of the surface water and replace it with fresh water);
- treating the pond water by **liming** (see Section 54) and/or **fertilization** (see Chapter 6) to increase phytoplankton only when it is necessary.

(d) **Reduction of oxygen requirements of fish by**:

- reducing the number of fish present in the pond;
- reducing or even stopping their supplementary feed.

26 How to improve water oxygenation at pond inlets

1. It is relatively easy to improve the oxygenation of the water **as it drops into the pond** (see **Pond construction, 20/2**). Several simple ways to do this are described below. You should select the system best adapted to your needs. **The mixing of atmospheric oxygen and water will improve as**:

- the height of the water drop increases;
- the width of the water and area of contact with the air increases;
- splashing and breaking of the water into fine droplets increases.

2. If the water supply is delivered **to the pond through a pipe**, you can improve oxygenation by:

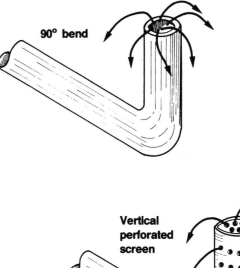

90° bend

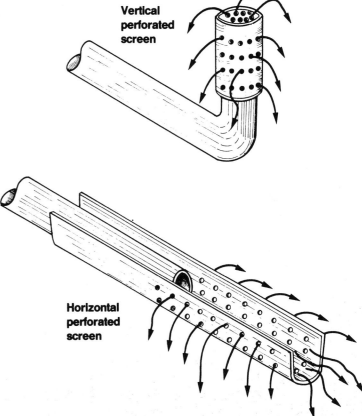

Vertical perforated screen

- adding **a 90° bend** at the end of the pipe and turning its opening up;
- fixing **a vertical perforated screen** over the upturned pipe end to increase the contact with the air;
- fixing **a horizontal perforated screen** so that it curves around the end of the pipe and sticks out in front of it.

Horizontal perforated screen

3. If the water supply **drops vertically into the pond** through any type of overhanging inlet, such as a pipe or a wooden frame, you can improve oxygenation with:

- **a horizontal splashboard**, fixed to a wooden support above the maximum water level;
- **a horizontal splashing and perforated device**, made of perforated metal or fine metal mesh fixed above water;
- **an inclined splashboard**, improved by transversal strips;

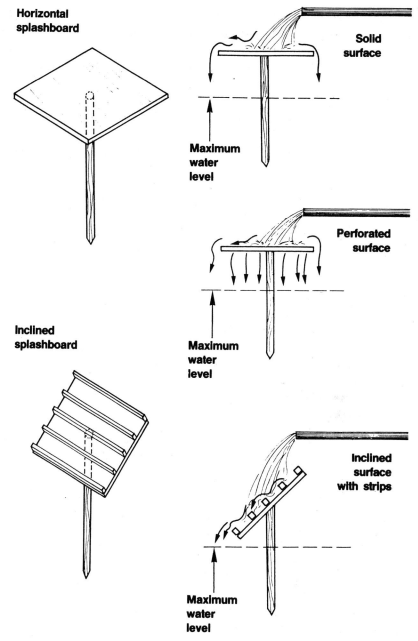

Horizontal splashboard

Solid surface

Maximum water level

Perforated surface

Maximum water level

Inclined splashboard

Inclined surface with strips

Maximum water level

41

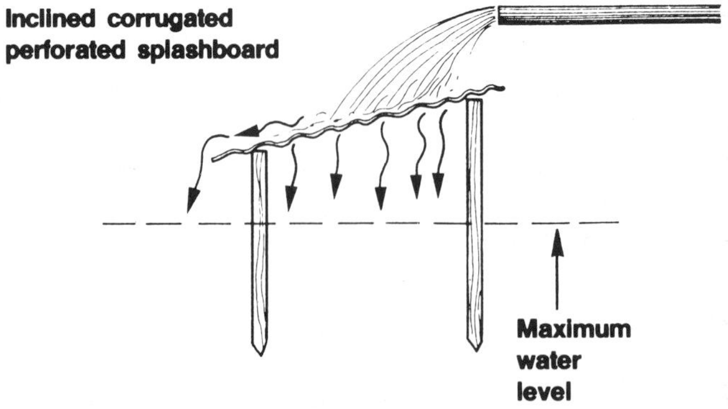

**Inclined corrugated
perforated splashboard**

**Maximum
water
level**

- an **inclined piece of corrugated iron** or asbestos sheet, preferably perforated with numerous 8-mm holes;
- a small **aeration column**, made of pipe or netting, with bits of
  plastic or netting inside;
- a **rotating water mill** made of wood;
- a **rotating water wheel**, also made of wood but suspended from
  the water inlet.

Rotating water wheel

**Maximum
water
level**

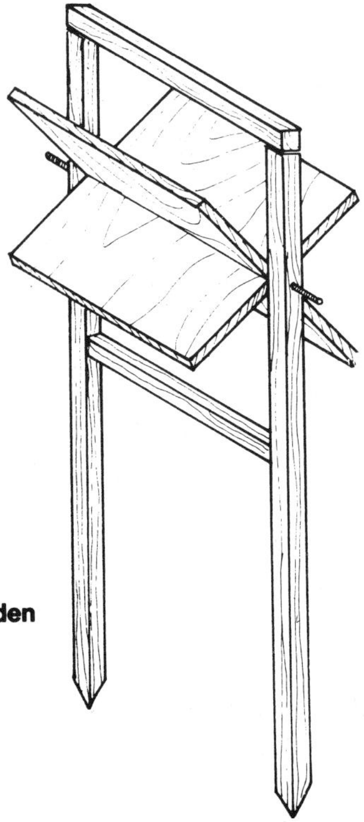

4.   If the water supply flows into the pond **through an inclined surface** at
the end of the feeding canal, the possibilities for improving its oxygenation are more limited. The most efficient improvement would be to place
a slanting piece of perforated **corrugated iron or asbestos sheet** above
the dike side.

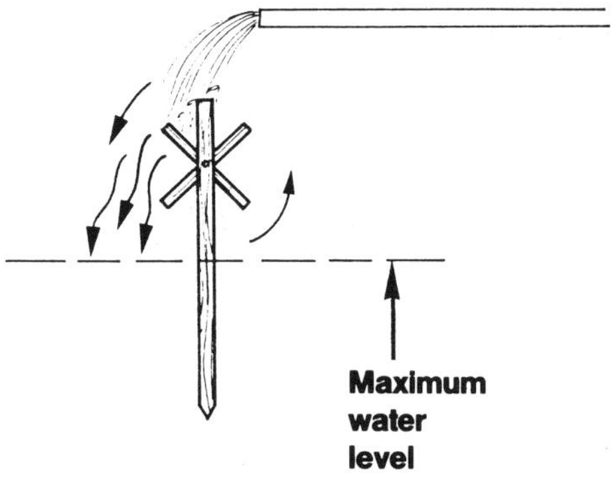

**Construction diagram
for building a simple wooden
rotating water wheel**

Note: to maintain their efficiency, you should regularly clean **meshed or
perforated** materials of any algal growth or other deposit.

27 How to improve oxygenation by using a cascade

Introduction

1. A cascade, which can be built using wood or stone or both, is a simple fixed structure over which **water is dropped**. As it drops, the water mixes with air, and its DO content is increased. This simple process is called **gravity aeration**, as opposed to the more complex mechanical aeration discussed later (see Section 28).

Using a cascade

2. A cascade is especially useful for increasing the DO content of **poorly oxygenated water**, for example spring water or water pumped from a deep well. It is also useful for oxygenating **water flowing out of a pond** before it is fed into another pond built in series (see **Pond construction, 20/1**). This step is particularly important for ponds stocked at high densities such as storage ponds and wintering ponds.

3. The use, position and design of a cascade depends on **the topography of the site**, particularly the available differences in elevation (see **Topography for freshwater fish culture**, *FAO Training Series*, **16/2**). Generally a cascade is built:

- **across a stream** supplying water to the fish farm, upstream from the main water intake, to increase the oxygenation of the supply water;
- **on the feeder canals** of ponds built in parallel, close to the ponds they service;
- **between ponds built in series**.

Note: if there is a risk of the water supply losing oxygen before it enters the ponds (e.g. if there is a lot of organic matter in the water), it is better to build a cascade near the pond entrance rather than at the beginning of the main water feeding canal.

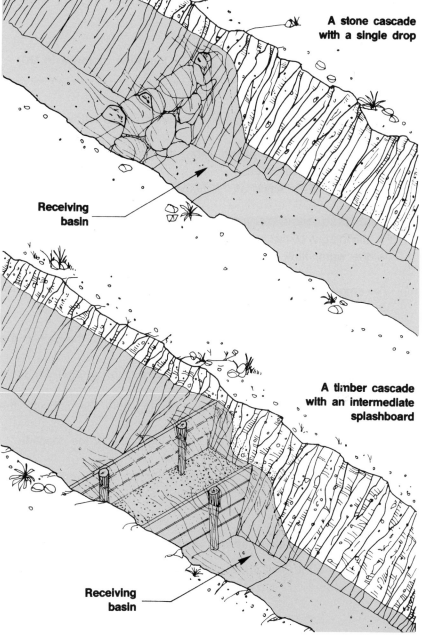

A stone cascade with a single drop

Receiving basin

A timber cascade with an intermediate splashboard

Receiving basin

43

Designing an efficient cascade

4. **The efficiency of transfer of atmospheric oxygen** into the water is usually defined by:

$$E = (100 \times \textbf{actual increase in DO}) \div (\textbf{possible increase in DO})$$

5. This transfer efficiency essentially depends on the following factors.

(a) **The initial DO content of the water**: the lower the DO content, the greater the improvement.

(b) **The energy of the water** as it falls over the cascade, which depends on:

  - the height of the cascade;
  - the water flow over the cascade.

(c) **The amount of mixing possible** between the water and the air, which depends on:

  - the number and nature of the devices placed in the path of the water;
  - the surface area of water exposed to air: the greater the area, the better the water oxygenation;
  - the depth of the flow of water running over the cascade: the shallower the better;
  - the depth and shape of the receiving basin of the cascade.

6. The greatest **efficiency** is, therefore, obtained with:

  - maximum height;
  - maximum width;
  - maximum breaking up and mixing of the water and air.

7. **When you plan a cascade**, take into account the following criteria, and incorporate the topographic features of the site.

(a) Give the cascade the **maximum height** possible at the particular location.

(b) If this **height is greater than 1 m**, break the cascade into two or three intermediate sections.

(c) If the **water drops from less than 1.40 m**, it is best to break the fall up across a horizontal surface, without using a receiving basin.

(d) If the **water drops from higher than 1.40 m**, it is best to break it up vertically, or as an incline, and to build a receiving basin with a water depth of at least one-tenth of the dropping height.

(e) Make the cascade wide enough so that the water depth over its top edge is less than 10 cm and preferably less than 5 cm. Try to break up the water flow at the edge by using a ridged bar, perforated screen, etc.

(f) You can increase efficiency **by using certain devices** which improve mixing, such as ridge bars, mesh screens or perforated sheets, but you should not forget to **clean them regularly** to maintain their efficiency.

Potential efficiency of different cascades for water oxygenation (in percent)

Type of cascade	Efficiency (E)	
	Total cascade height (cm)	
	30.5	61.0
Simple	9.3	12.4
One splashboard at mid-height	24.1	38.1
Inclined corrugated sheet down to mid-height	30.1	43.0
Same with holes	30.1	50.1

28 Mechanical aeration devices

1.   Mechanical aeration devices are relatively expensive and require an **external power source**. They are consequently most commonly used for short-term and emergency aeration, such as converting extreme daytime oxygen depletion when ponds are particularly heavily stocked, or during extreme climatic periods. In more intensive fish farming, mechanical aerators may be used continuously, but this step is only practical for higher-value fish species under extremely well-managed and controlled conditions.

Choosing devices

2.   All of the devices mentioned on this page and on page 46 are used **inside the pond**. They are powered either directly, with small electric or gasoline motors, or indirectly (e.g. from a tractor power take-off), or using a shore-based pump or air compressor. Some common forms of mechanical aeration include the following:

(a)   **Spray-type (fountain) aerators**: these usually have an electric motor that spins a vertical propeller, which drives water upwards through a spreader device, spraying water out in an arc around the aerator. The aerator is mounted on a float, which is tethered in position in the pond. A simple guard protects the fish from the spinning propeller. These aerators are typically 0.25 to 1 kW in size.

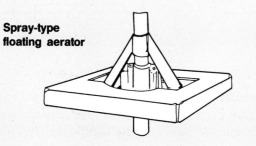

Spray-type floating aerator

(b)   **Pumps**: the water is pumped out from a pond, or an adjacent water body, and sprayed back on to the pond surface. Pumps can also be used to feed water to a simple **cascade**. Pumps can be powered by electricity, gasoline or a diesel engine, or from a tractor.

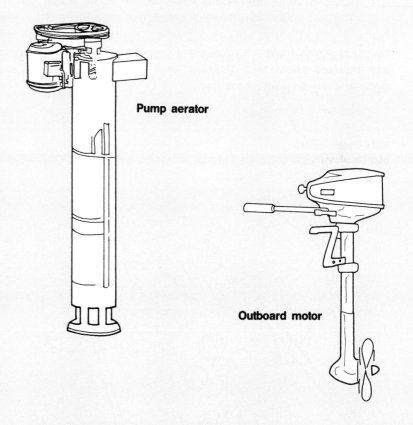

Pump aerator

Outboard motor

(c)   **Outboard motors**: to be used to mix and aerate pond water, they should be attached to a fixed point such as a wooden platform or a securely moored boat. Care should be taken to protect the fish from the propeller.

(d) **Paddle-wheel aerators**: a float-mounted motor drives a horizontal axle on which is fitted at each end a vertical paddle-wheel, set about one-fourth to one-third of its diameter down into the water. The blades of the paddle-wheel are usually slightly scooped or angled and are perforated. The paddle-wheels rotate in the water, scooping, lifting and splashing the water out across the surface. They are typically 0.5 to 2 kW in size. Paddle-wheel aerators can also be set up using power from a tractor, via a propeller shaft. Some models are mounted on wheels, allowing them to be backed up into the pond and towed away after use.

Commercial paddle-wheel aerator

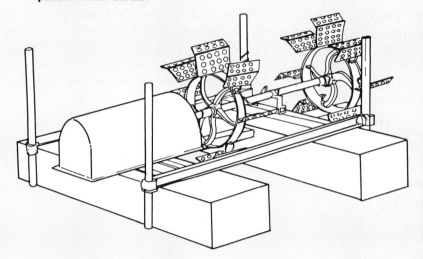

Homemade metal paddle-wheel aerator

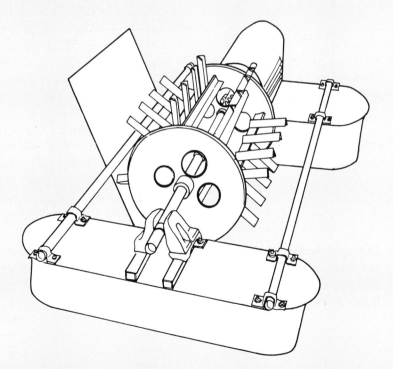

Tractor-driven paddle-wheel aerator

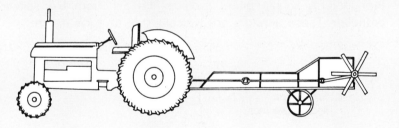

(e) **Air compressors or blowers**: these can be used together with a perforated pipe or air diffusers. They are more common in smaller ponds or for special uses such as harvesting. The air compressor feeds compressed air down the pipe, from which it emerges in the form of fine to medium-sized bubbles, which mix with the water to provide aeration. These systems are better used in deeper water.

Selecting mechanical aerators

3. Mechanical aerators are usually defined by efficiency and capacity.

4. **Efficiency** is usually expressed as kilogram (kg) of oxygen transferred per kilowatt-hour (kWh) of energy applied. This figure typically ranges from 0.2 to 1.5. As with cascades, the level depends on factors such as the amount of mixing created and the existing DO level in the water.

5. **Capacity** is usually expressed as kg of oxygen transferred per hour. It varies with the size and power of the device, but typically ranges from 0.5 to 5 kg/h.

6. As an appropriate guide, it is best to allow for **at least 5 kg/h per hectare of pond**, corresponding to an increase of DO concentration of about **0.5 mg/l per hour**. At a typical efficiency of 1 kg/kWh, this increase would require 5kW (about 6.7 HP) of applied power per hectare.

7. **Selecting the type and size** of mechanical aerators depends on several factors, including:

- the availability and local cost of the aerator;
- the size of the unit and its flexibility of use (e.g. to move from pond to pond, or to reposition within a pond);
- whether you prefer to use several smaller units or a single larger unit, the former being more flexible, but more expensive;
- whether you wish to mix and circulate the pond water, seeing that some types, such as paddle-wheels, are better at mixing and moving the water than others.

29 Water filtering devices

Introduction

1. Good fish farm management requires proper control over water supplies and their quality. **Water filtering devices** are commonly used:

- **to improve water quality** by reducing turbidity and removing some of the suspended organic materials such as plant debris (see also Section 70, **Pond construction, 20/2**);
- **to limit the introduction of wild fish**, which can compete for food, transmit infections and diseases and reduce the farm's production; the more carnivorous species can kill your stock, particularly the smaller fish;
- **to keep farmed fish** from escaping from the pond.

2. There are several types of water filter, from very simple devices suitable for small rural fish farms, to more elaborate filters better adapted to the larger water flows or special needs of commercial farms. **The location** of these devices within the farm may vary. Some are set up on individual ponds, while others can service the whole fish farm. Some are specifically used for pond inlets. **Table 2** on page 48 will help you **choose the most appropriate filtering device**.

3. Water filtering devices should be **properly maintained** to retain both their filtering efficiency and their water discharge capacity. **Clean them regularly** according to the water turbidity and the amount of material collected. Check them and **repair them** as soon as the filtering efficiency alters. **Remember** that the smaller the particle size your filter can retain, the more susceptible it becomes to clogging, and the more frequently you should check it and clean it.

TABLE 2

Water filtering devices

Water filtering device	Location				Recommended						Build it yourself		Paragraphs
	Feeder canal	Tanks	Pond inlet	Pond outlet	Pipes	Grooved structures (e.g. monks)	Other structures	Water flow			Free materials	Bought materials	
								Small	Medium	Large			
Cover screens	—	—	●	●	●	—	—	●	—	—	●	●	4-5
Sleeve filters	—	●	●	—	●	—	—	●	●	●	—	●	6-12
Fence screens	●	—	●	●	●	—	●	●	●	●	●	●	13-16
Fence filters	—	—	●	—	●	—	●	●	●	—	●	—	17-18
Sliding screens	—	●	●	●	—	●	—	●	●	—	—	●	19-21
Horizontal screens	●	—	—	—	—	—	—	●	●	—	—	●	22-28
Box/basket filters	—	—	●	—	●	—	●	●	●	●	●	●	29-31
Barrage filters	●	—	—	—	—	—	—	—	●	●	●	●	32-35
Reverse filters	—	●	—	—	●	—	—	●	●	—	—	●	36-38

Cover screens for water pipes

4. Pond inlet and outlet pipes can be cheaply screened **by covering their end** with some flexible filtering material. This material should **fit tightly** over the pipe end and be **well secured**, if necessary using a piece of nylon string or galvanized wire.

5. Depending on how much money you wish to spend, you can, for example, use one of the following:

- a perforated tin can;
- a funnel made of split/woven bamboo;
- a cap made of wire or plastic mesh.

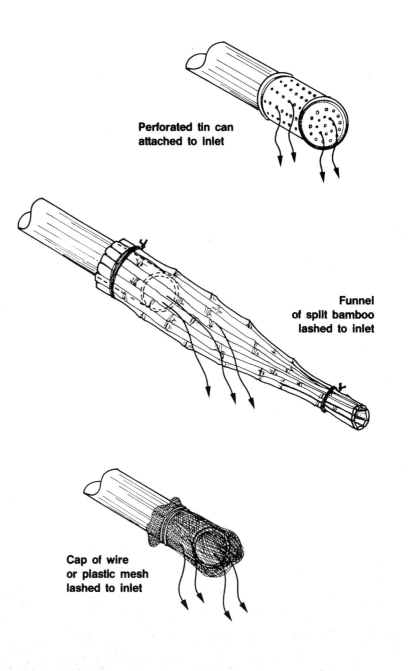

Perforated tin can attached to inlet

Funnel of split bamboo lashed to inlet

Cap of wire or plastic mesh lashed to inlet

6.   **A sleeve filter** consists of a **cylindrical mesh tube**, securely fastened around the inlet pipe at one end and tightly closed with a string at the lower end.

7.   It is a simple and relatively cheap device, which allows the passage of a large volume of water while **very efficiently retaining** even the eggs and larvae of wild fish. However, as it is usually made of very fine-meshed synthetic material (e.g. saran with 21 meshes per cm), it may **clog up rapidly** if its overall surface area is not large enough to handle the water turbidity and flow.

8.   If you have the right synthetic material, you can easily **make a sleeve filter yourself** in the following way.

(a)   Cut two rectangular pieces of mesh material with the following dimensions, depending on the size you need:

- from 40 to 90 cm wide; and
- from 1.5 to 4 m long.

Note: the width of the piece of mesh to be cut should be equal to $(1.57 \times D)$ + $(4 \times S)$ where **D** is the diameter (in cm) of the sleeve filter to be made and **S** is the width (in cm) of the double flat seam joining together the two pieces of mesh.

(b)   Fold the edges in two and using strong thread, such as nylon, sew the pieces together with a double row of stitches and two double flat seams, to make a cylinder.

(c)   Fit each end of this cylinder with a drawstring closure and nylon string.

Note: a saran sleeve filter 4 m long and about 50 cm in diameter should be sufficient to handle 60 l/s water flow. You need about one square metre of filter surface per 10 l/s water flow.

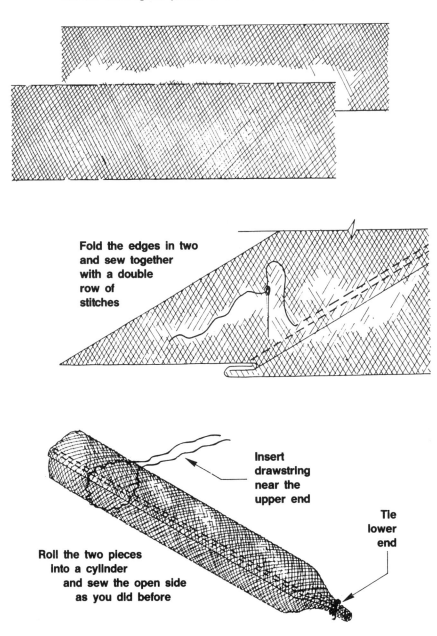

Cut two rectangular pieces of mesh

Fold the edges in two and sew together with a double row of stitches

Roll the two pieces into a cylinder and sew the open side as you did before

Insert drawstring near the upper end

Tie lower end

9. How the sleeve filter is used depends on the vertical distance between the inlet pipe and the pond water surface.

(a) **If this distance is quite large**, for example when filling an empty pond, it is best to fit the sleeve filter inside a wooden frame to support the whole of its length.

(b) **If this distance is relatively small**, for example in a fully filled pond, the sleeve filter will last longer if it floats in the water.

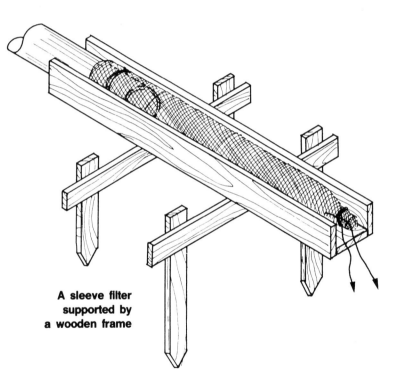

A sleeve filter supported by a wooden frame

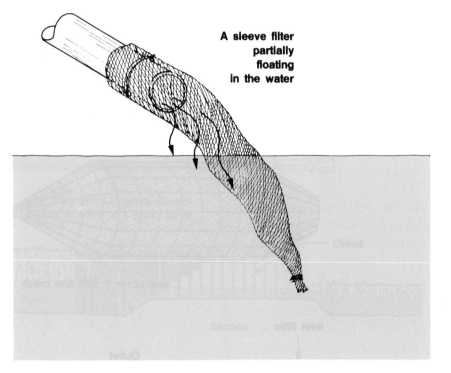

A sleeve filter partially floating in the water

10. Secure one end of the sleeve filter around the end of the inlet pipe. For additional security, **attach the end of the string to a strong anchoring point**.

11. Tightly close the other end of the sleeve filter with its string. Make sure that it can be easily opened later, simplifying cleaning.

51

Fence filters for pond inlets

17. Fence filters are somewhat similar to fence screens, except that they are much thicker and heavier, and are **built mainly with wood and stones**. To build a cheap, durable structure with locally available materials, proceed as follows.

(a) In front of the pond inlet, using string and pegs, mark **a semi-circular area** centred on the axis of the inlet at a distance of at least 1 m. The area itself should be about 40 cm wide.

(b) Along the inside and outside limits of this area, drive **two vertical rows** of wooden poles solidly into the earth about 40 cm apart. The poles should stick about 30 cm out of the ground.

(c) Remove 15 to 20 cm of bottom soil from the area.

(d) Build **a solid foundation** by laying rows of big stones near the poles, within the cleared area.

(e) Tightly fill the spaces in between with smaller stone chips. Do not leave large holes in the inner part of the filter.

(f) Build the filter up, successively adding layers of bigger stones (on the outside, near the poles) and small stone chips (on the inside), until you reach slightly below the top of the poles.

18. If the inflow water transports too much debris or silt, the filter may eventually become clogged. At least once a year, take it apart for cleaning and rebuild it properly.

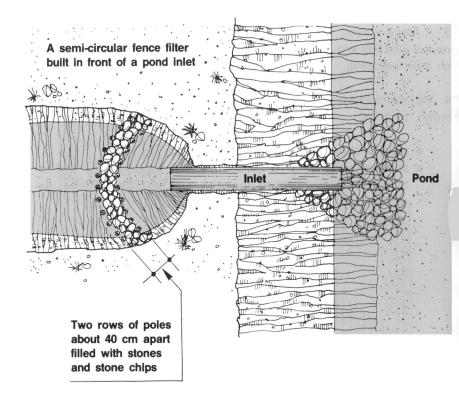

A semi-circular fence filter built in front of a pond inlet

Inlet

Pond

Two rows of poles about 40 cm apart filled with stones and stone chips

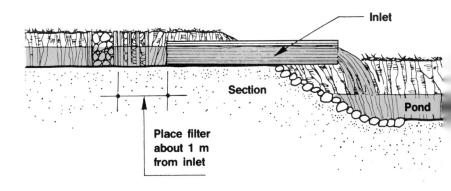

Inlet

Section

Pond

Place filter about 1 m from inlet

Sliding screens for grooved structures

19. In a previous manual (see **Pond construction**, **20/2**), you learned how most water transport and control structures are built with **sets of grooves** that hold wooden boards and sliding screens. Such filtering screens are commonly used in main water intakes, at pond inlets, in monk and sluice outlets and in holding tanks as well as in structures such as those used for harvesting and grading fish.

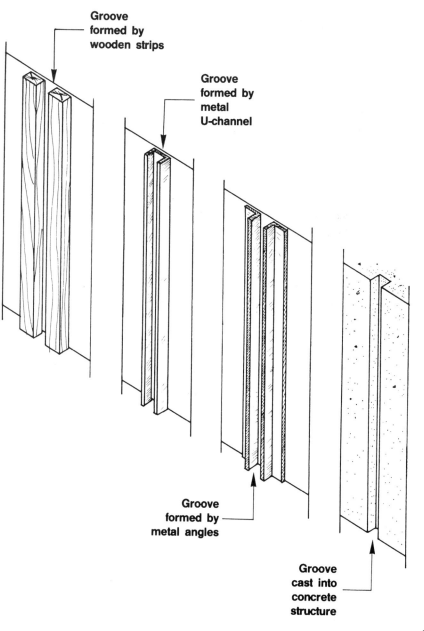

Groove formed by wooden strips

Groove formed by metal U-channel

Groove formed by metal angles

Groove cast into concrete structure

20. Sliding screens usually consist of **an external frame** to which **the filtering material** is attached. They may be designed according to the following criteria.

(a) **The dimensions**, depending on the structure for which they are built:

- **the width** should be about 1 to 2 cm less than the width of the structure (i.e. groove to groove), so that the screen can easily slide up and down;
- **the height** should be calculated to allow at least 5 to 10 cm clearance above the maximum water level in the canal or in the pond;
- **the working area** of screens can be only part of the height, for example to fit on top of a column of wooden boards, or the full height, for example to filter the whole water column.

(b) **The frame material**, generally either wood or metal: the wood should be either selected for durability under water, or appropriately treated (see Section 31, **Pond construction, 20/1**).

(c) **The filtering material**, generally either metal or plastic mesh. Round iron bars welded to a metal frame may also be used.

(d) **The filtering efficiency**, the greater the efficiency, the smaller the holes needed and the more susceptible the screen is to clogging. To reduce the need for very frequent cleaning, use larger filtering areas for fine screens. Do not use holes that are too small.

21. As an approximate guide, **the velocity of water** through the open area of a screen should not exceed 0.02 m/s. Thus, if **the expected flow** is 10 l/s (0.01 m³/s), **an open area** of at least 0.01 m³/s ÷ 0.02 m/s = 0.5 m², is required. If the filter screen has 20 percent open area in normal operation, you will need a 2.5-m² screen.

Note: if your farm has several ponds, it will be too time-consuming to look after a fine-mesh screen at the inlet of each pond; you should either build a good general filtering device upstream from your ponds (see next section), or you should use box/basket filters (see paragraphs 29 to 31).

56

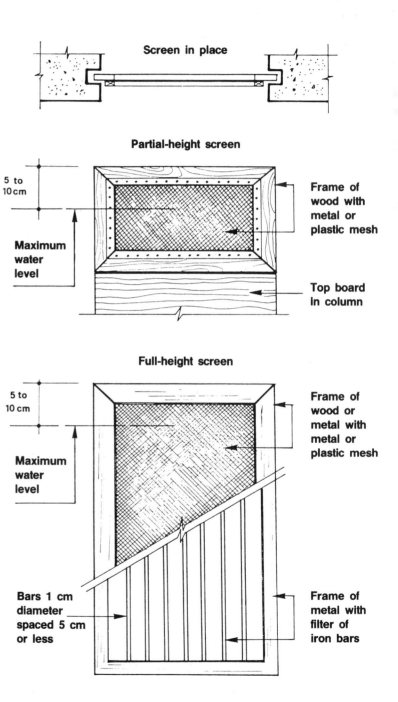

Screen in place

Partial-height screen

5 to 10 cm

Maximum water level

Frame of wood with metal or plastic mesh

Top board in column

Full-height screen

5 to 10 cm

Maximum water level

Frame of wood or metal with metal or plastic mesh

Bars 1 cm diameter spaced 5 cm or less

Frame of metal with filter of iron bars

22. **The horizontal underwater screen** is a simple and efficient filtering device built at the beginning of the main feeder canal, which services the whole fish farm. It provides **continuous filtering of the water supply** with little risk of clogging with floating debris. It can be cheaply built, if necessary with the assistance of a builder.

(a) On the basis of the maximum water flow required for the fish farm, estimate how large the **surface area of the horizontal screen** should be (see paragraph 21). Divide this area by the width of the feeder canal at the surface of the water to find **the screen length** necessary.

(b) In the main feeder canal, **first build a rectangular open box** using either wood, bricks, cement blocks or concrete. Build the box as follows:

- **its inside width** should equal the width of the canal at the water surface;
- **its length** should be at least 50 cm more than the calculated screen length;
- **its inside depth** should equal the depth of the canal (thus **the floor level** should be at the level of the canal bottom);
- at the front and back of the box, build in one pair of **grooves**;
- along each side of the box, for a little more than the length of the screen, fix wooden or metal **brackets to support** the horizontal screen, setting the top of these brackets at a horizontal level about 15 cm below the usual level of the water surface in the canal.

23. **The filtering device** is to be tightly fitted inside the box between the two pairs of grooves, and is built as follows:

(a) Build a rectangular **wooden, plastic or metal frame** on which you fit **the filtering screen**. This screen should be made of strong material, preferably a perforated metal sheet, but plastic or metal mesh may also be used. Mesh size should not be smaller than 3 mm.

(b) At the back (downstream) end of this frame and **perpendicular** to it, attach **a wooden board**. Its height should be equal to the distance from the top of the support brackets to the top of the box built in the canal.

(c) At the front (upstream) end of the filtering frame and **perpendicular** to it but hanging down, attach **with hinges a wooden board**. Its height should be slightly more than the distance from the top of the support brackets to the bottom of the box built in the canal.

(d) **Position this filtering device in the box** on the supporting brackets, with the hinged board turned down and facing the entrance of the water canal. **Check and adjust** as necessary in order that:

- the filtering device fits close enough to the side walls of the box;
- the screen frame sits level on the brackets;
- the front hinged board fits tightly against the bottom of the box;
- the back vertical board reaches near to the top of the box.

24. The drawings on pages 58 and 59 show two examples of horizontal underwater screens, one in wood and one in metal, and how to place them in a feeder canal.

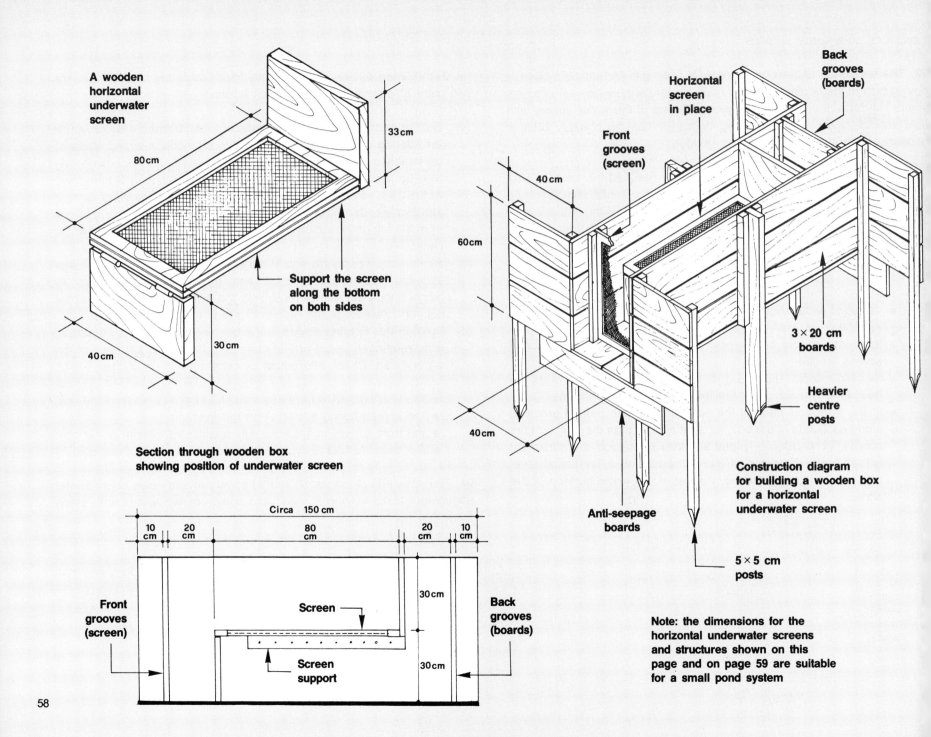

A wooden horizontal underwater screen

80 cm

33 cm

Support the screen along the bottom on both sides

40 cm

30 cm

Section through wooden box showing position of underwater screen

Circa 150 cm

10 cm

20 cm

80 cm

20 cm

10 cm

Front grooves (screen)

Screen

Screen support

30 cm

30 cm

Back grooves (boards)

Horizontal screen in place

Back grooves (boards)

Front grooves (screen)

40 cm

60 cm

40 cm

3 × 20 cm boards

Heavier centre posts

Anti-seepage boards

Construction diagram for building a wooden box for a horizontal underwater screen

5 × 5 cm posts

Note: the dimensions for the horizontal underwater screens and structures shown on this page and on page 59 are suitable for a small pond system

58

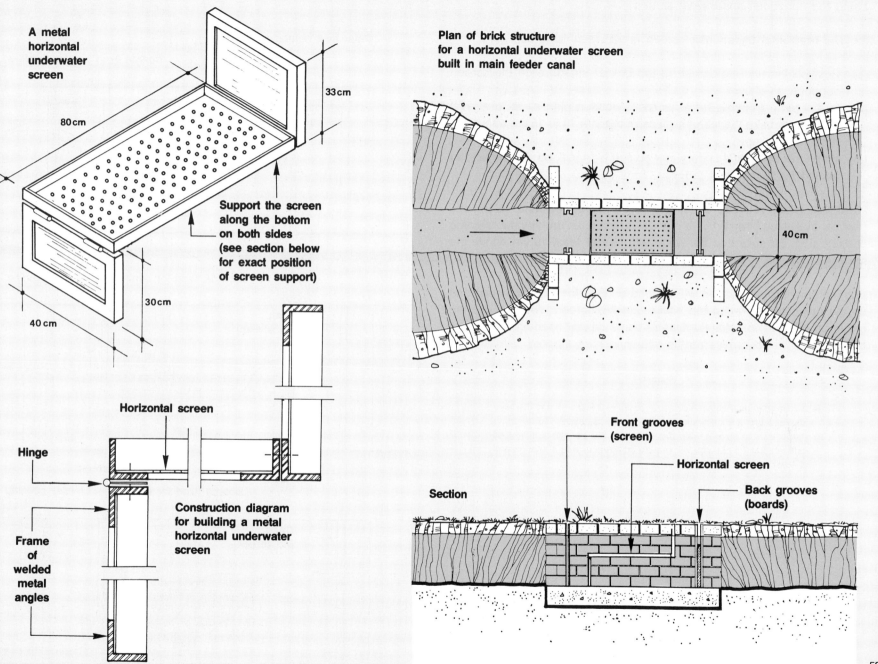

A metal horizontal underwater screen

80 cm

33 cm

30 cm

40 cm

Support the screen along the bottom on both sides (see section below for exact position of screen support)

Horizontal screen

Hinge

Frame of welded metal angles

Construction diagram for building a metal horizontal underwater screen

Plan of brick structure for a horizontal underwater screen built in main feeder canal

40 cm

Front grooves (screen)

Horizontal screen

Back grooves (boards)

Section

59

Using the horizontal underwater screen

25. **The horizontal underwater screen** is very simple to use.

(a) Position the filtering screen in the box or structure on the screen supports as shown.
(b) **In the back grooves**, set a series of wooden planks, up to a level which will make the water surface rise about 10 to 15 cm above the horizontal screen.
(c) When the horizontal underwater screen is in use, **no vertical screen should be in the front grooves**.

26. If you wish, you can easily **regulate the flow of water** along the channel by setting boards in the grooves to make an **underwater opening** of fixed size. Be sure that the water will rise **at least 10 cm** above the horizontal screen.

27. The flow-through system will depend on **the size of the opening** and on **the head loss** between the intake and the screen. You can use the graph on page 61 to estimate flow for some typical opening sizes. Note that the flow will decrease as the filter becomes blocked.

Example

- If your opening measures 60 cm × 30 cm = 1 800 cm^2, and the head loss is 5 cm, the maximum water flow will be about 110 l/s.
- If you want to have a water flow of about 300 l/s with a head loss estimated at 7.5 cm, you will require an opening with a total area of at least 4 150 cm^2.

Water flow regulation

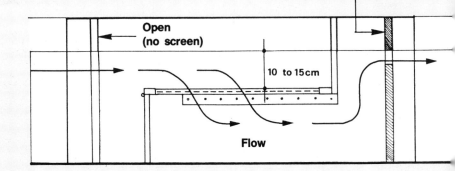

Boards fixed for top overflow

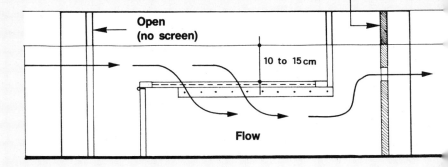

Boards fixed for centre underwater opening

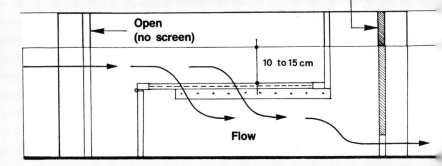

Boards fixed for bottom underwater opening

GRAPH 2

Water flow according to opening size and head loss

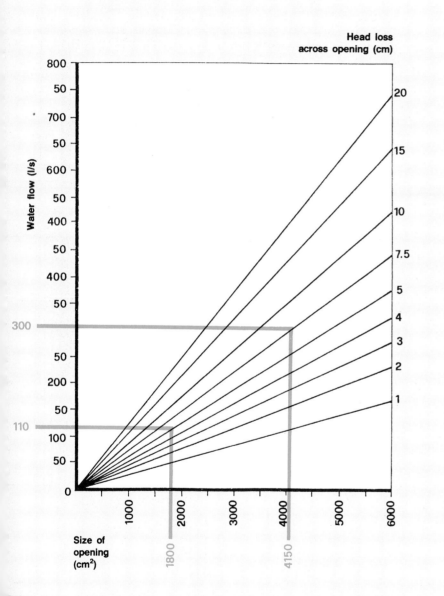

Head loss
across opening (cm)

Water flow (l/s)

Size of
opening
(cm²)

28. **To clean the screen**, proceed as follows.

(a) Insert a full-height sliding screen in the front grooves.
(b) By hand remove all the floating debris accumulated in front of the vertical board at the back end of the screen.
(c) Grab the top of this board and lift the horizontal screen partly out of the water. The front vertical board on which the screen is hinged should remain in its position underwater.

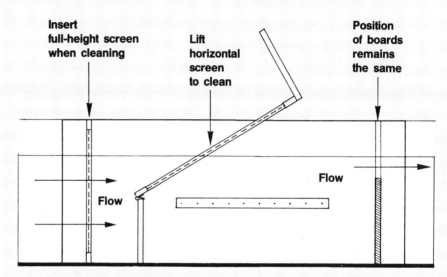

Insert
full-height screen
when cleaning

Lift
horizontal
screen
to clean

Position
of boards
remains
the same

Flow

Flow

(d) Brush the screen well.
(e) Set the screen back underwater on its supports.
(f) Remove the sliding screen from the front grooves.

29. **A box or basket filter** consists of a filtering container which is placed under **the pond inlet**. Debris and undesirable fish fall with the inflowing water into the container, where they remain trapped while the water filters through. Different sizes of material can be trapped depending on the efficiency of the filter used.

30. Ideally, access to and **cleaning of the filters** should be easy. Although their price is often higher than sliding inlet screens, box or basket filters have **some advantages**:

- they require less maintenance;
- they may be more efficient; and
- they can handle a larger flow of water.

31. There are several kinds of box or basket filters. Some are easy to build yourself with locally available materials. For others, you will need to buy special filtering cloth and may require the assistance of a good carpenter. Here are some examples.

(a) **Wooden box filter attached to inlet pipe**: a strong filtering screen should be used at the bottom of the box. It is useful for small water flows and rural ponds.

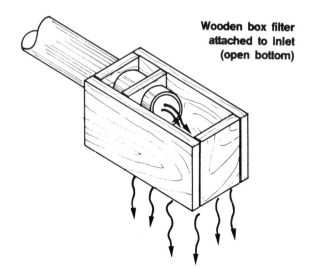

**Wooden box filter
attached to inlet
(open bottom)**

(b) **Screened box filter under inlet**: the force of the inflowing water is first broken on the strong bottom. Mesh material is only used on the sides. Four legs support the box filter under the inlet at a level above maximum water level in the pond. This type is more expensive to build, and dimensions depend on water inflow: for 100- to 1 000-m^2 ponds, a box of 50 × 50 cm with 20 to 40 cm height is suitable.

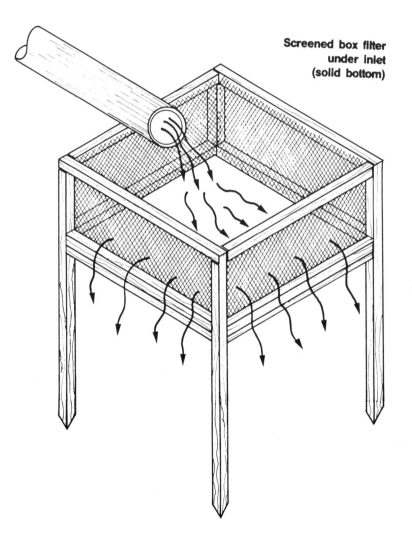

**Screened box filter
under inlet
(solid bottom)**

(c) **Basket filter filled with stones or chips**: about 0.5 m above the pond water surface, fix under the inlet a wooden basket or half a metal drum, with bottom and sides fully perforated. Fill the outer sides with larger stones. Partly fill the middle portion with stone chips. In the upper part, leave a 20-cm hollow in which the water drops. Check and clean the basket regularly to prevent it from getting clogged.

(d) **Perforated metal drum under inlet**: an old metal drum, open at the top, is perforated with holes on the bottom half of its sides. Use a strong nail from the outside of the drum. Fix the drum under the inlet so that the inflowing water drops into the middle of it. Filtration efficiency depends on the size of the holes.

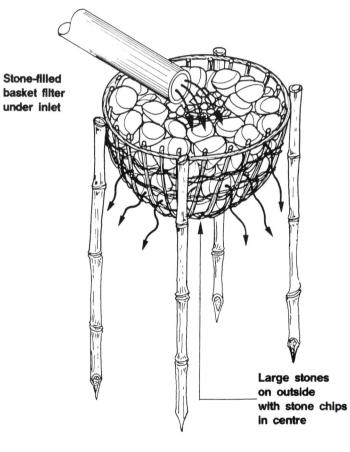

Stone-filled
basket filter
under inlet

Large stones
on outside
with stone chips
in centre

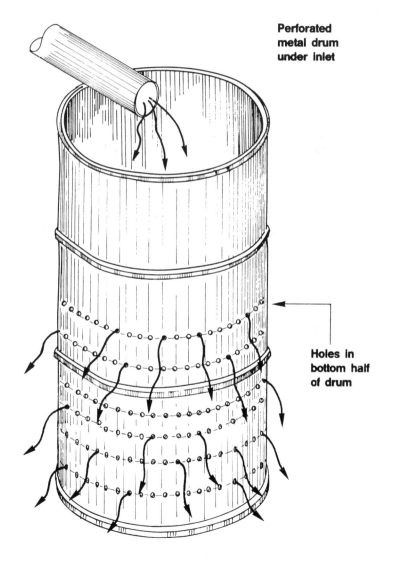

Perforated
metal drum
under inlet

Holes in
bottom half
of drum

63

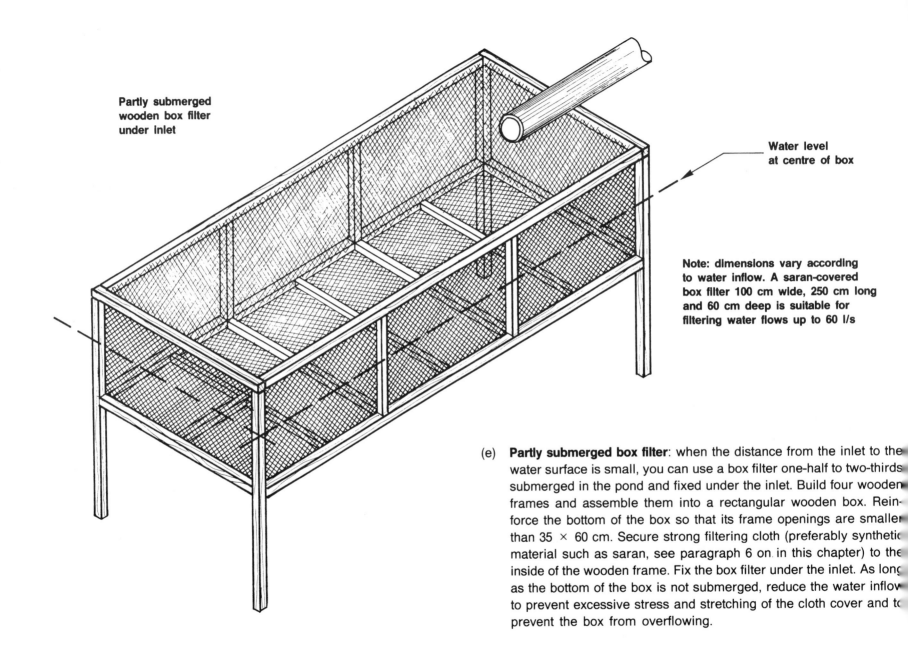

Partly submerged wooden box filter under inlet

Water level at centre of box

Note: dimensions vary according to water inflow. A saran-covered box filter 100 cm wide, 250 cm long and 60 cm deep is suitable for filtering water flows up to 60 l/s

(e) **Partly submerged box filter**: when the distance from the inlet to the water surface is small, you can use a box filter one-half to two-thirds submerged in the pond and fixed under the inlet. Build four wooden frames and assemble them into a rectangular wooden box. Reinforce the bottom of the box so that its frame openings are smaller than 35 × 60 cm. Secure strong filtering cloth (preferably synthetic material such as saran, see paragraph 6 on in this chapter) to the inside of the wooden frame. Fix the box filter under the inlet. As long as the bottom of the box is not submerged, reduce the water inflow to prevent excessive stress and stretching of the cloth cover and to prevent the box from overflowing.

(f) **Floating box filter**: similar to the previous model, except that it floats in the pond, under the inlet, with the bottom part well submerged. This filter is good in greater water depths or with varying water levels. It is preferred to the previous type whenever the inlet is high enough above the pond surface:

- **a small to large box filter** can be made as described above. Build a wooden-framed box, reinforced at the bottom. Secure the filtering cloth to the inside of the frame. Fasten some floats, such as small empty plastic drums, at each corner of the box filter so as to keep it afloat with about one-half to two-thirds of its depth submerged. Place it under the water inlet and maintain it in that position with four corner-ropes and anchoring points. Reduce the water flow if the bottom cloth is not well submerged, when filling a pond for example;

- **a small floating filter** can be made of a filtering cloth bag secured at its top to an old inner-tube. To keep it well under water, add some weight to the bottom of the bag, such as stones attached to each corner or a metal ring laid inside.

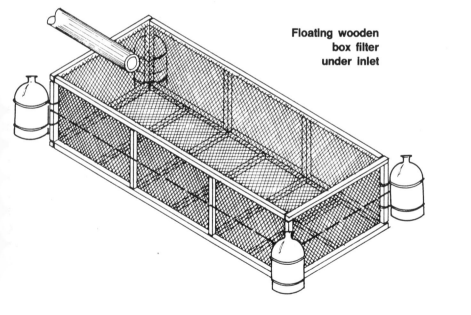

Floating wooden box filter under inlet

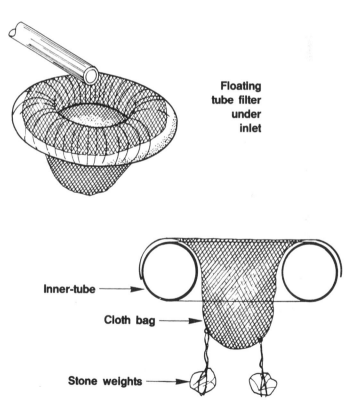

Floating tube filter under inlet

Inner-tube ⟶

Cloth bag ⟶

Stone weights ⟶

Barrage filters for feeder canals

32. **A barrage filter** is a fixed structure built on the feeder canal, generally at its beginning. It uses **stones, gravel and stone chips** as filtering material. When the filter becomes clogged with debris and silt, it should be removed, washed and placed back in position. The structure can be made simply and inexpensively if only earth, wood and gravel, all locally available, are used. It becomes more expensive if bricks or concrete are used to build a chamber and its inflow/outflow portions.

Designing barrage filters

33. **A barrage filter** acts as a barrier in the feeder canal, and it can considerably **reduce the flow of water** unless it has a large enough area and a sufficient head to push the water through it. When designing the dimensions of your barrage filter, you should allow at **least 1 m³ of filter for every 1 l/s** of water flow, and allow **at least 5 cm of head** to permit water flow through the filter. The filter should be **at least 1 m thick**.

34. As a general rule, make the filter **at least four times as wide** as the feeder canal on which it is to be built.

Building barrage filters

35. Here are some examples of how you can build a barrage filter on your main feeder canal.

(a) **A simple and inexpensive barrage filter** can be designed as a small, shallow pond measuring 20 × 25 m. Between two parallel rows (distance 1.5 m) of wooden poles (diameter 3 to 12 cm) driven 30 cm vertically into the ground at close intervals, stones, gravel and stone chips are piled up to a level slightly higher than the maximum water level in the canal, for example 1 m.

How to widen a feeder canal for a barrage filter

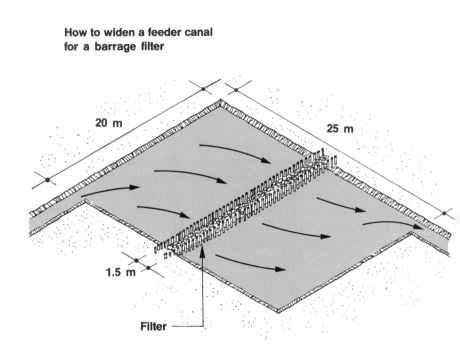

20 m

25 m

1.5 m

Filter

Section through stone- and gravel-filled filter

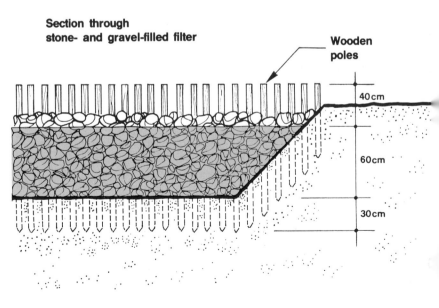

Wooden poles

40 cm

60 cm

30 cm

Concrete block barrage filter
(20 × 20 × 40 cm blocks)

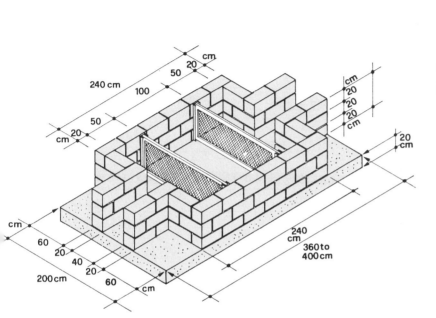

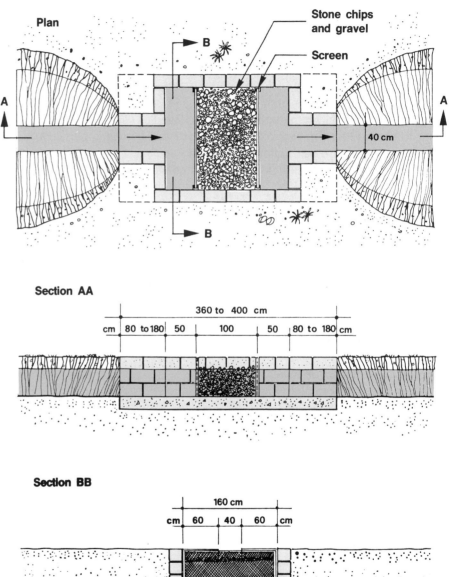

Plan

Stone chips and gravel

Screen

40 cm

Section AA

360 to 400 cm

cm | 80 to 180 | 50 | 100 | 50 | 80 to 180 | cm

Section BB

160 cm

cm | 60 | 40 | 60 | cm

b) **A barrage filter with a brick or block chamber**: the feeder canal is enlarged to build a rectangular chamber on a solid foundation (see **Pond construction, 20/1 and 2**, in this series). Across this chamber two metal screens made, for example, of heavy metal mesh or round iron bars, are set in grooves 1 m apart. Stone chips and gravel are piled up tightly between the screens, to slightly above the maximum water level in the canal. The structure is completed with an inlet and an outlet channel, also of brick or block, from 80 to 100 cm in length.

A somewhat **more sophisticated barrage filter**, similar to the one shown on this page, can also be built in concrete with 7.5-cm walls and a 10- to 15-cm foundation.

67

36. **Upflow filters** are more sophisticated and are more expensive to build. Skilled masonry work is essential for their construction. The filter needs to be set up and handled very carefully or it may fail to work. Repairs are generally difficult and costly. However, such a filter has the advantage of making it possible **to reverse the water flow** (backwash) and to wash out routinely the deposited silt and other particles from the filtering material. Such reverse cleaning should be carried out regularly, sometimes even daily if the turbidity is high. Regular cleaning allows the filter to be used for removal of quite fine particles. Its use is particularly appropriate where clean water is essential, such as in a hatchery for breeding fish (see Sections 90 to 94, **Management, 21/2**).

37. **To operate an upflow filter**, proceed as follows.

(a) Be sure **the water inflow passes through a screen first** to remove the larger debris and avoid clogging the draining valve. Check that the draining valve is closed.

(b) Open the **inlet valve** and **the outlet valve**. The water should pass through the filtering material under some pressure.

38. **To clean the upflow filter**:

- close the inlet valve;
- open the draining valve;
- establish a reverse flow through the open outlet valve to run water through the filtering material from top to bottom and clean it;
- close the draining valve when the water flowing out of it is clean.

The filter is then ready to be used again.

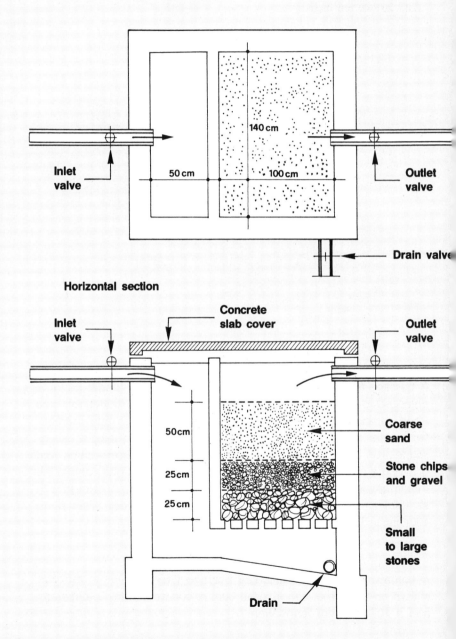

Plan of masonry upflow filter

Horizontal section

3 CONTROLLING WATER LOSSES IN PONDS

1. In a previous manual in this series (**Soil, 6**), you learned how important it is to select a site with **suitable soil** for a fish farm. A good choice makes it possible to build ponds with strong impervious dikes and with relatively impervious pond bottoms (see also **Pond construction, 20/1**).

The selected site

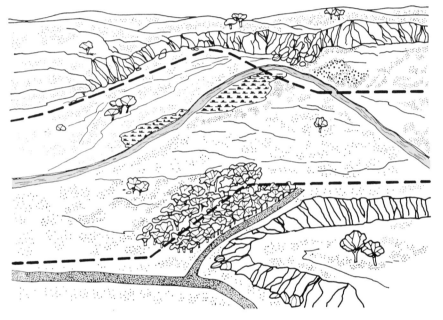

2. It may happen that your fish pond is not as good as it should be. **Losing much water through seepage**, for example as much as 10 cm per day (see Section 21, **Water, 4**), may be attributable to one of the following:

- **the site**, although known as less than satisfactory, had to be used;
- the **soil survey** was not done well;
- excessive **removal of surface soil** for dike construction has exposed highly pervious areas of sand, gravel or rock;
- the pond was not well designed or constructed.

3. As well as losing water, excess seepage results in a continuing loss of fertility. You should aim to **reduce seepage** to a point at which water losses become at least tolerable, for example **less than 5 cm per day**.

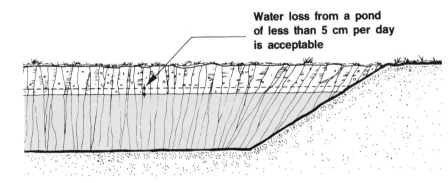

Water loss from a pond of less than 5 cm per day is acceptable

4. **A newly built pond usually loses more water than an older pond**. As the pond is being used and well managed, the organic matter produced from fertilization and feeding falls to the pond bottom, where it gradually blocks the soil pores and so reduces the bottom soil permeability.

Note: if seepage losses are due to bad design or construction, e.g. lateral seepage through dikes which are too narrow, the losses can be very difficult to stop.

5. There are several methods you can use:

- either **accelerate at minimal cost** the **natural process** of sealing the soil pores; or
- if this is not possible, seal the pond bottom using **additional material** at extra cost.

6. You will learn more about reducing seepage losses in the next sections. Use **Table 3** to select a suitable method according to the quality of the soil to be sealed and the local availability of materials.

7. In the final section of this chapter, you will learn how to repair leaking ponds during periodic maintenance.

TABLE 3

Control of excessive water seepage in ponds (according to soil texture)

Pond sealing method		Soil texture	Silt content	Clay content
Accelerating the natural process				
Cattle manure	31	Variable but less	More than 50%	
Gleization	31	than 50% sand		
Changing soil structure				
Puddling	32	Wide range of particle	Present	More than 10%
Compaction	33	sizes from sand to clay		
Lining the pond				
Soil blanket	34	Mostly coarse particles	Very little	Less than 10%
Synthetic membrane	35			
Adding high-swelling clay				
Bentonite	36	—	—	—

31 How to seal the pond with organic matter

Light organic layer

1. A simple method that greatly accelerates the sealing of a pond bottom **is to spread enough cattle manure over it** (at least 2 m³/100 m²) before filling the pond with water. Repeat this procedure several times if necessary, at intervals of eight to 12 months, until seepage losses become acceptable. After draining these ponds, refill them with water as soon as possible to reduce the rate of decomposition of the organic matter that seals the bottom pores.

2. If the dikes of a pond have been compacted very well, it may not be necessary to seal them against seepage when you seal the pond bottom. However, if you are not sure that the dikes are sound, it is best to seal them as well.

Gleization

3. If the bottom soil is too permeable to be sealed by this method, an alternative is to create **an impervious biological plastic layer** in the bottom and on the sides of the pond. Such an impervious layer is called **a gley**, and the process of its formation is called **gleization**. Proceed as follows.

(a) Prepare the pond bottom (and if necessary the wet sides of the dikes) by clearing it of all vegetation, sticks, stones, rocks and the like. Fill all cracks, crevices and holes with well-compacted impervious soil.

(b) Completely cover the cleaned surface with **moist animal manure**, preferably pig dung, spread in an even layer about 10 cm thick.

(c) Cover the manure well with a layer of **vegetal material**, preferably broad leaves such as banana leaves. You can also use dried grass, rice straw, soaked cardboard or paper, etc.

(d) Cover with a layer of soil about 10 cm thick.

(e) Moisten and compact very well.

(f) Wait two to three weeks before slowly filling up the pond with water.

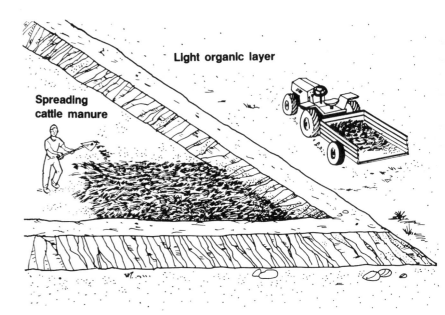

Light organic layer

Spreading cattle manure

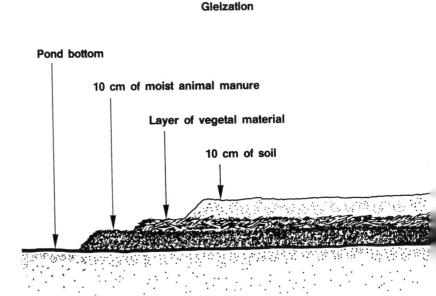

Gleization

Pond bottom

10 cm of moist animal manure

Layer of vegetal material

10 cm of soil

2 How to seal the pond bottom by puddling

A simple way to reduce water seepage, particularly if the pond bottom is very dry, hard and has open cracks in it, is **to break the soil structure** of the pond bottom before filling the pond with water. This is common practice in irrigated rice fields, and is called **puddling**. Refer also to Section 21, **Water, 4**. Proceed as follows.

a) Saturate the soil of the pond bottom with water.
b) Let the water soak into the soil just enough to permit working.
c) Break the soil structure by puddling with a hoe or plough.

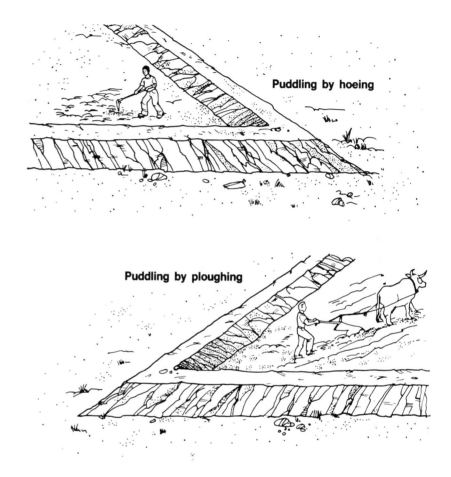

Puddling by hoeing

Puddling by ploughing

If **a puddler** is available, flood the pond bottom and **puddle the soil under water** for best results. A plough can also be used underwater.

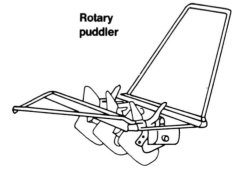

Rotary puddler

33 How to seal the pond bottom by compacting

1. If the bottom soil contains **a wide range of particle sizes**, from coarse sand to fine silt and clay, and **at least 10 percent clay**, it can be made relatively impervious by good **compaction** alone (see **Soil, 6**). Proceed as follows.

(a) Prepare the pond bottom (and if necessary the wet sides of the dikes) by clearing it of all vegetation, sticks, stones, rocks and the like. Fill all cracks, crevices and holes with well-compacted impervious soil.

(b) **Loosen the soil to a depth of 20 to 25 cm**. If possible, try to borrow a disc harrow or a rototiller from a neighbouring farmer. A plough could also be used.

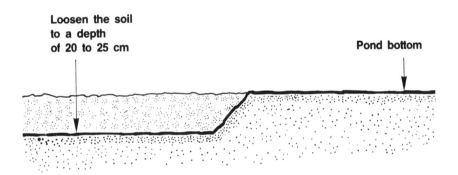

(c) As necessary, dry or moisten the loosened soil so that it reaches its **optimum moisture content for compaction**, about 13 percent (see Section 102, **Soil, 6**).

(d) **Compact the soil surface** into a dense, tight layer. You should preferably use four to six passes of a sheepsfoot roller. You may also use the tyres of a heavy tractor or a crawler-tractor.

(e) **Check compaction** carefully. If necessary, readjust the soil moisture content and improve compaction. Refer also to Section 62, **Pond construction, 20/1**.

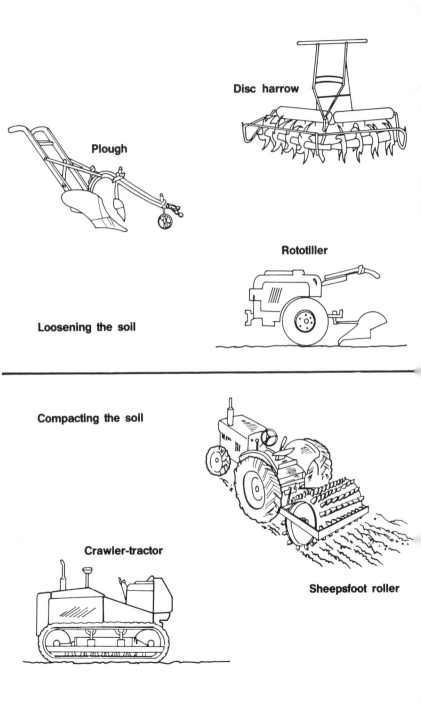

Plough

Disc harrow

Loosening the soil

Rototiller

Compacting the soil

Crawler-tractor

Sheepsfoot roller

34 How to seal the pond bottom with a soil blanket

Laying out a soil blanket

1. If the bottom soil contains a great proportion of coarse particles **but lacks enough clay and silt** to prevent excessive seepage, you can seal it with **a soil blanket**. Proceed as follows.

a) Prepare the pond bottom (and if necessary the wet sides of the dikes) by clearing it of all vegetation, sticks, stones, rock and the like. Fill all cracks, crevices and holes with well-compacted impervious soil.

b) Build a soil layer **about 15 cm thick** over the entire bottom area and over the wet sides of the dikes. The soil you use should be **well graded and contain at least 20 percent clay**.

c) Moisten this soil layer so as to reach **optimum moisture content for compaction** (see Section 102, **Soil, 6**).

d) **Compact** the soil layer **well**, preferably by four to six passes of a sheepsfoot roller (see Section 62, **Pond construction, 20/1**).

e) Bring in additional clay soil **to build a second layer about 15 cm thick** over the first one.

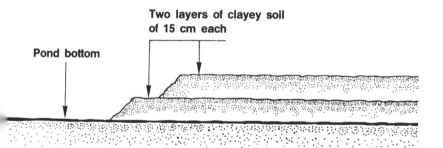

Two layers of clayey soil of 15 cm each

Pond bottom

Moisten and compact as before.
Protect the soil blanket against rupture by:

- using layers of rocks or stones at the **water inlet**;
- spreading a cover of gravel 30 to 45 cm thick just below the maximum water level, if there is a danger of water freezing;
- avoiding working inside the pond over the soil blanket when the pond is drained or fish are being harvested;

- not letting the soil blanket dry too much and thus cracking;
- refilling the pond carefully as soon as possible.

Note: try to obtain the clay soil required to build the blanket from a **borrow area close enough to the pond** to reduce transport costs.

Laying out a subsurface soil blanket

2. For smaller ponds or if you need to work inside the pond, you may choose to use a method which provides **better protection for the soil blanket**. Proceed as follows.

(a) Prepare the pond bottom (and if necessary the wet sides of the dikes) by clearing it of all vegetation, sticks, stones, rocks and the like.

(b) **Excavate the pond bottom** deeper to a depth of about 30 cm. Store the soil close to the pond.

(c) Build **a well-compacted layer of clayey soil** at least 5 cm thick all over the surface.

(d) To protect this layer, cover it with the original soil and compact the soil cover.

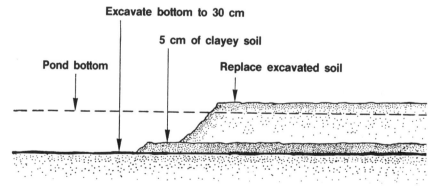

Excavate bottom to 30 cm

5 cm of clayey soil

Pond bottom

Replace excavated soil

Note: if seepage problems are suspected during planning or construction, it would be cheaper to include this subsurface blanket during pond construction. This step will make it easier **to connect properly with the clay core** of the dikes.

TABLE 4

Relative properties and requirements of synthetic membranes

Properties	Thermo-plastic membranes		Elastomer membranes
	Polyethylene (PE)	Polyvinyl chloride (PVC)	Butyl rubber
Relative cost	Small	Medium	High
Ageing	Good	Not so good	Good
Resistance to impact	Low	Medium	High
Susceptibility to sunlight		Poor to medium[1]	Medium to good
Protection required			
Normal use	Soil layer at least 15 cm thick		None
Trampling by livestock, people, equipment	Soil layer at least 25 cm thick with bottom 7.5 cm very fine sand[2]		Soil layer 20-25 cm thick with bottom 7.5 cm very fine sand
Joining or patching	Heat, special cement, or tape	Solvent cement	Special cement
Minimum membrane thickness			
over sands	0.20 mm or 8 mils	0.20 mm or 8 mils	0.38 mm or 15 mils
over gravels	0.38 mm or 15 mils	0.38 mm or 15 mils	0.76 mm or 30 mils
Placement in pond	Lay with 10% slack	Lay smooth but slack	Lay smooth but slack

[1] During manufacture, can be treated for increased resistance to sunlight
[2] In some cases, a layer of geotentile is used to give additional strength. This is a loose weave synthetic fibre mesh which helps hold together the lower layer of soil and acts as a backing to the membrane. It is laid the same way as the membrane itself

35 How to seal the pond with a synthetic membrane

1. Thin synthetic membranes can also be used to reduce excessive seepage, particularly in coarse-grained soils. Such a method has **the advantages** of:

- reducing seepage to zero in all types of soil; and
- providing a more dependable seal than the previous method.

2. However, synthetic membranes have **several disadvantages** which must be carefully considered.

(a) They are rather expensive.
(b) Their resistance to puncture and breakage is limited.
(c) This resistance may be weakened under the action of direct sunlight.
(d) The natural fertility of the bottom soil will be lost, and it will take some time to develop it over the membrane.

3. Three kinds of synthetic membrane material are most commonly used:

- black polyethylene;
- polyvinyl chloride (PVC); or
- butyl rubber.

4. Their relative properties and general requirements when being used to line fish ponds are summarized in **Table 4**.

Note: the thickness of synthetic membranes is expressed either in **millimetres** (mm) or in **thousandths of an inch** (mils). To convert one unit to the other, simply remember that **1 mil = 0.0254 mm**, or approximately 40 mils = 1 mm.

5. To ensure zero seepage, the synthetic membrane must entirely cover the pond bottom and the wet sides of the dikes. Since it is relatively easy to puncture the membrane, it is especially important to clean both the bottom and the wet sides of the dikes very well.

6. **To line a pond with a synthetic membrane**, proceed as follows.

(a) Prepare the entire pond area (including the wet sides of the dikes) by clearing it of all vegetation, sticks, stones, rocks and the like. Fill all cracks, crevices and holes with well-compacted impervious soil.
(b) Fill holes and crevices with soil.
(c) If the soil material to be lined is **too stony or of very coarse texture**, cover it with a cushion layer of finer material about 10 cm thick.
(d) All around the pond on the top of the dikes, dig **an anchor trench** 25 cm deep and about 30 cm wide, set at least 30 cm back from the inner edge of the dike.
(e) **Lay a first strip** of the membrane along the width of the pond and along the dike. Watch for the following:

- choosing the correct slack for the membrane used (see **Table 4**);
- leaving 20 to 25 cm of the membrane in the bottom of the anchor trench at both ends;

Position of synthetic membrane ending in anchor trenches

Fold the membrane into the trench as shown

Synthetic membrane

Anchor trench

Anchor trench

- avoiding soil on top of the membrane;
- avoiding puncturing the membrane while handling it.

(f) **Lay a second strip** of the membrane in the same way but allowing **a 15-cm overlap** for joining it to the first strip.

(g) **Carefully join this second strip** to the first one, using the appropriate method according to the membrane used (see **Table 4**).

(h) **Repeat this process** with as many membrane strips as necessary to completely cover the pond bottom and the wet sides of the dikes.

(i) Backfill the anchoring trenches with well-graded soil and **bury the edges** of the membrane, taking care to compact well.

(j) While laying the membrane strips down, **cover the finished sections for immediate protection** by:

- first carefully spreading and compacting a 7.5-cm-thick layer of soil no coarser than silty sand (**Soil, 6**), and being very careful to avoid puncturing the membrane;
- then spreading and compacting a layer of available site soil at least 15 cm thick.

(k) In some cases **the membrane is not covered**. If so, you should use a covering sheet of plastic, butyl or felt over the upper edge that extends at least 20 cm below the pond water level to protect the main membrane from heat and sunlight. The sheet can be dug into the same trench as the main membrane.

(l) You should take **special care around pipes or monks**. Depending on the membrane type, it can be sealed either by taping or cementing around the pipe or monk. With monks, a finer mortar joint can also be used: ideally the membrane should be incorporated during construction of the monk. Bring the membrane at least 10 cm up the sides of the monk or around the pipe, and make sure the membrane is not split around the joint.

Note: if the synthetic membrane is to be covered with a protective soil layer, the slope of the wet side should be no steeper than 3:1; if there is no soil cover, the slope of the wet side should be no steeper than 2:1.

78

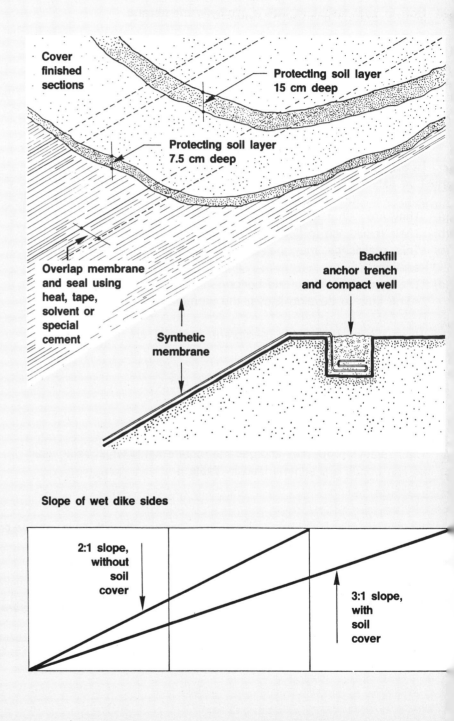

Cover finished sections

Protecting soil layer 15 cm deep

Protecting soil layer 7.5 cm deep

Backfill anchor trench and compact well

Overlap membrane and seal using heat, tape, solvent or special cement

Synthetic membrane

Slope of wet dike sides

2:1 slope, without soil cover

3:1 slope, with soil cover

1. If your pond bottom soil contains a high percentage of coarse-grained particles and not enough clay, adding **bentonite** to it is another method of reducing excessive seepage.

What is bentonite?

2. Bentonite is a fine-textured colloidal clay with as much as **90 percent of montmorillonite** (see Section 18, **Soil, 6**). When exposed to water, dried bentonite absorbs several times its own weight of water; at complete saturation, it swells as much as eight to twenty times its original dry volume.

3. **Natural bentonite** deposits exist in several places around the world, for example in the United States of America, Mexico, China and Western Europe.

Selecting this method

4. Before selecting this method to seal your pond, you should locate **a satisfactory source of bentonite**, as near as possible to your site to reduce cost.

5. You should also ensure that the quality of the available bentonite is good: it should contain **the highest percentage of montmorillonite** possible (at least 70 percent). If necessary, obtain a sample first and have it analysed in a soil laboratory. You may test the bentonite by placing it in a small jar and adding water. Good bentonite should expand to several times its original volume.

6. **The water level** of your pond should not fluctuate too much. Soil treated with bentonite will develop **numerous cracks** when it dries up. It is best if it **always remains wet**.

7. Whenever good bentonite is readily available, this sealing method offers **several advantages**:

- it is cheaper than using some synthetic membranes such as butyl rubber;
- it is relatively easy to apply;
- it can offer a long-lasting solution to excessive seepage;
- unlike plastic membranes, it is not vulnerable to vandalism, it does not stretch and break if the substrate sags under pressure, and vegetation can grow through it without affecting its porosity.

8. However, bentonite treatment has **several potential disadvantages**:

- cost may be excessive if the transport distance is too long;
- it is more laborious to apply than a butyl membrane, and great care should be taken to ensure complete coverage;
- it can be disrupted by cattle or eroded by running water unless well protected;
- burrowing animals such as crayfish or crabs and strong bottom-feeders such as common carp can penetrate through the bentonite-treated layer;
- bentonite treatment is not advisable in highly alkaline soils;
- it should be used with care when the water contains salts such as calcium chloride.

Taking precautions before applying the method

9. Before using this method, it is safer first to make **some preliminary tests**.

(a) Have a **water sample** analysed to discover its salt content.
(b) Have **soil samples** analysed for chemical composition and texture.
(c) Request **specialized advice from your supplier** of bentonite: which particular quality of bentonite do you require and at which rate should you apply it.

10. If the results of the above **analyses of water and soil** confirm that conditions are good for using a particular kind of bentonite to seal your pond, proceed as follows.

(a) Prepare the pond bottom (and if necessary the wet sides of the dikes) by clearing it of all vegetation, sticks, stones, rocks and the like. Fill all cracks, crevices and holes with well-compacted impervious soil.

(b) Dry or moisten the area so that the **soil moisture** reaches its **optimum content for compaction** (see Section 102, **Soil, 6**).

(c) According to the bentonite and soil laboratory analyses, **uniformly spread from 5 to 15 kg of bentonite per m^2**: the better the quality of bentonite and the more clay or silt your soil contains, the less bentonite you will require. Use a marked grid pattern to help you spread the bentonite on the bottom evenly.

(d) Thoroughly **mix the bentonite with the bottom soil** to a depth of about 15 cm. A rototiller is best for this operation but an agricultural disc harrow can also be used.

(e) **Compact the area well**, preferably with four to six passes of a sheepsfoot roller. You could also use other types of compacting equipment (see Section 62, **Pond construction, 20/1**).

(f) If you cannot fill the pond with water immediately, you should **protect the treated area** against drying and cracking. Use a good mulch* of straw and hay on top of the surface and keep it moist if necessary.

(g) **Protect areas** where erosion could damage the layer containing bentonite such as under the water inlet and around the pond where wave action is thought to be the strongest (see Section 43). Use gravel, stones, rocks and other available materials.

Note: when you plan to use bentonite to seal a pond and its **wet dike sides, the dike slope ratio should be no steeper than 3:1**.

Sealing a full pond with bentonite

11. If you discover that seepage losses from a pond in use are excessive and bentonite is easily available, **you can seal the pond without draining it**. First check on this possibility by having water and soil samples tested.

12. At the pond inlet, **throw bentonite in the inflowing water** at the rate of 5 to 15 kg/m^2 of water surface, according to the results of the bentonite and bottom soil analyses. You may also **sprinkle the bentonite on the entire water surface**. The bentonite particles will settle down on the pond bottom and penetrate the soil pores as the water seeps through. As they swell, they will block these pores and reduce seepage losses.

Pond bottom

Layer of bentonite

Mix bentonite and bottom soil to a depth of 15 cm

Note: it is best to treat your pond with bentonite in the absence of fish, because it is potentially harmful to their gills.

37 How to maintain earthen ponds

1. An earthen fish pond needs to be given most attention **during the first years of operation**, because it is then that the unforeseen, hidden faults come to the fore. Maintenance should therefore be carried out continuously from the beginning (see Chapter 16, **Management, 21/2**).

2. **With earthen ponds and in particular with dikes**, you should expect several kinds of problems, which you should take care of as soon as possible.

(a) **Surface erosion of the dikes** is usually caused by rain trickling down the side slopes. Protect these slopes with a healthy cover of grass (see Section 43).

(b) **Erosion of the wet side of the dikes** at the water surface level mainly attributable to strong wave action may justify additional protection (see Section 43).

(c) **Part of the soil may slip** down from the slope in some places, most frequently resulting from bad compaction, drenching of the dike or quick water discharge. Replace the slipped soil with **well-compacted horizontal layers** of soil.

(d) **Part of the dike may slump**, resulting either from bad compaction or from incomplete removal of organic matter from the dike's site. Loosen the surface soil of the dike top, moisten to optimum moisture content for compaction, add a layer of good quality soil and moisten again if necessary and compact well.

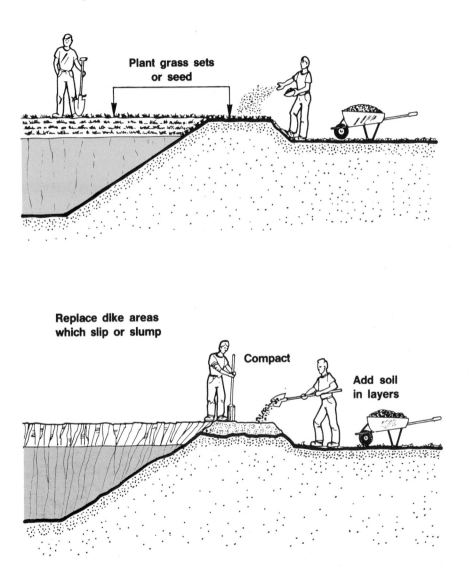

Protect dikes with grass cover

Plant grass sets or seed

Replace dike areas which slip or slump

Compact

Add soil in layers

3. **If a dike breaks at one point** and a gap develops, **the water level must be lowered** immediately to a level below the gap.

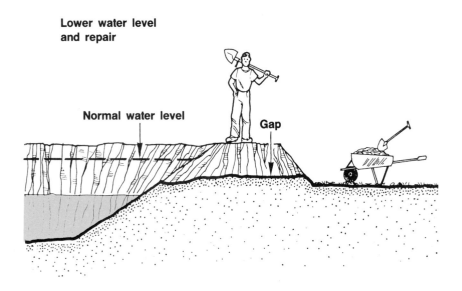

Lower water level and repair

Normal water level

Gap

4. Then carefully repair the dike as follows.

(a) **Widen and deepen the gap slightly** to expose fresh and clean dike material. Eliminate all organic matter that might be present, such as roots. Make irregular cuts to give the repair a good foundation.

(b) Moisten the old dike material, and pack in layer after layer of good quality new soil. As you build up the new dike section, **give particular attention to the junction of the old soil with the new soil**.

(c) If **a clay core** is present, rebuild it also as you progress upwards, layer by layer (see **Pond construction, 20/1**).

5. **Leaks may develop locally in dikes** for various reasons, such as accidental damage, the presence of a rotting piece of root or the burrow of an animal. To repair local leaks, you can proceed in the following way.

(a) **Lower the water level**, slightly widen the leak and fill it with well-compacted clayey soil. You can also **use a mixture of clay and hydrated lime** (see Section 46), for example in the proportion of 4 to 6:1, which will set more firmly. In some areas, hydraulic limes, made from clayey limestone* (10 to 30 percent clay) can also be applied directly.

(b) If the leak is not too large, throw a certain amount of clayey soil into the water directly above the leak. You can also use sifted cinders (less than 2 mm in diameter) or a 5:1 mixture of finely grained soil with bentonite (see Section 36).

(c) For emergency use, you may also apply a section of polythene sheet, well weighted down with soil.

6. In tropical countries, particularly during the dry season, termites may establish themselves in the dike of a pond. The dike becomes weakened by underground galleries, which may result in severe water leaks and even breakages. To eliminate termites, proceed as follows.

(a) Partly open the anthill at the top.
(b) Pour used motor oil into the galleries. If available, use xylophene, because it has a stronger effect.
(c) Repeat this treatment each day for one week.
(d) Close the galleries with clayey soil, moistening as necessary and compacting well.

4   PROTECTING FARM STRUCTURES AND FISH STOCKS

40 Introduction

1. **Soil erosion has negative effects** on water quality and on the fish farm itself. You must protect your farm against erosion and its effects if you wish to maintain good levels of production at a reasonable cost. **Soil erosion** should be controlled, not only on the farm itself but also in the entire **catchment basin** upstream from it (see Section 16, **Water, 4**). This is usually the responsibility of the regional administration. All fish farmers involved should seek action or assistance from them if needed. However, on your own land and maybe even in its immediate neighbourhood, you should practise **soil conservation** wherever necessary. It is best if local farmers work together to achieve this goal.

**A stream network
with six catchment basins**

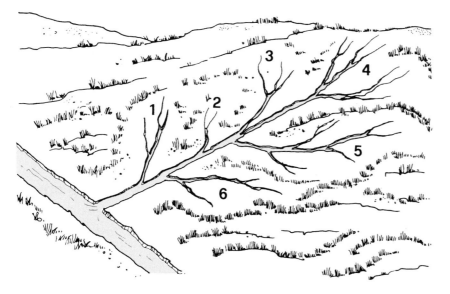

**Soil conservation practices
may be just as important on neighbouring lands
as on your own land**

2. Soil conservation practices include:

- **protecting sloping land** surrounding your fish ponds against rainfall erosion (see Section 41);
- protecting your fish ponds from **excessive winds** by establishing wind-breaks (see Section 42);
- protecting your fish ponds against eroding **wave action** (see Section 43); and
- protecting your pond dikes against **damage by rain** (see Section 44) and uncontrolled **livestock traffic** (see Section 45).

3. Needless to say, you should also be concerned to control erosion and its effects downstream of your farm.

4. **Pest control** in fish ponds applies both to harmful animals that may feed on your fish and to plants that may harbour minute organisms responsible for diseases and infections. Animals may compete with your fish for food, and both plants and animals may reduce the production potential of your ponds.

5. For best results, **animal pests** should be controlled in fish ponds:

- **after each total fish harvest and before restocking**: for ponds that are completely drained (see Section 46), and for ponds that are not completely drained (see Section 47); and
- **during the fish production cycle** (see Section 48).

6. **Vegetation control** can be carried out concurrently (see Section 49

41 Soil conservation

Introduction

1. You have already learned in a previous manual in this series (see **Water, 4**), that whenever the ground is not perfectly horizontal, **rainwater** partly infiltrates it and partly runs off over its surface. As water runs down a slope, it transports particles of the surface soil. The more water passes and the faster it runs, the bigger the particles that can be washed down the slope. This process is called **erosion**, and it can cause:

- serious degradation of the slope itself and of the soil properties, reducing fertility;
- an inflow of turbid water at the bottom of the slope and problems of soil settling elsewhere.

2. If your fish farm is situated at the bottom of a valley or even if your main water feeder canal runs across sloping ground, you should try to control soil erosion on the slopes as much as possible to prevent turbid water from running into your ponds. This practice, called **soil conservation**, can result in great benefits:

- **better soils** on the slopes and greater production of various products such as wood, fruits, fodder or food for you and your family;
- **better water quality** in your fish ponds and greater fish production.

Note: if you cannot control soil erosion in your farm watershed, you can use **a protection canal** to collect and divert turbid water. For example round a barrage pond or a sunken pond fed by runoff (see Section 115, **Pond construction, 20/2**). You can also improve the quality of your water supply by using **a settling basin** (see Section 116, **Pond construction, 20/2**).

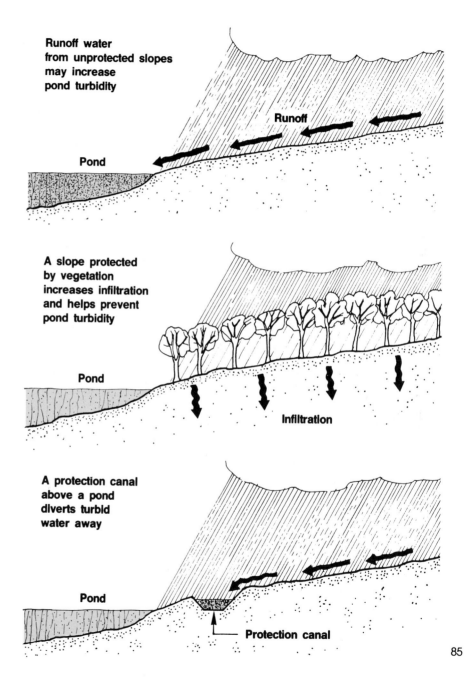

Runoff water from unprotected slopes may increase pond turbidity

Runoff

Pond

A slope protected by vegetation increases infiltration and helps prevent pond turbidity

Pond

Infiltration

A protection canal above a pond diverts turbid water away

Pond

Protection canal

Kinds of erosion

3. **The erosion of soil by rain** takes place in successive phases.

4. At the beginning, the rain hits the soil surface, causing soil particles to break off and burst upwards, while water progressively humidifies the soil surface and infiltrates deeper. This is called **splash erosion**. It loosens soil but does not move it much.

5. On flat land, as soon as the upper layer soil becomes saturated and infiltration decreases, a water layer forms over the soil surface and splash erosion stops; if the ground is not horizontal, water starts to run off the slope, transporting fine soil particles with it. This is called **sheet erosion**.

Splash erosion

Sheet erosion

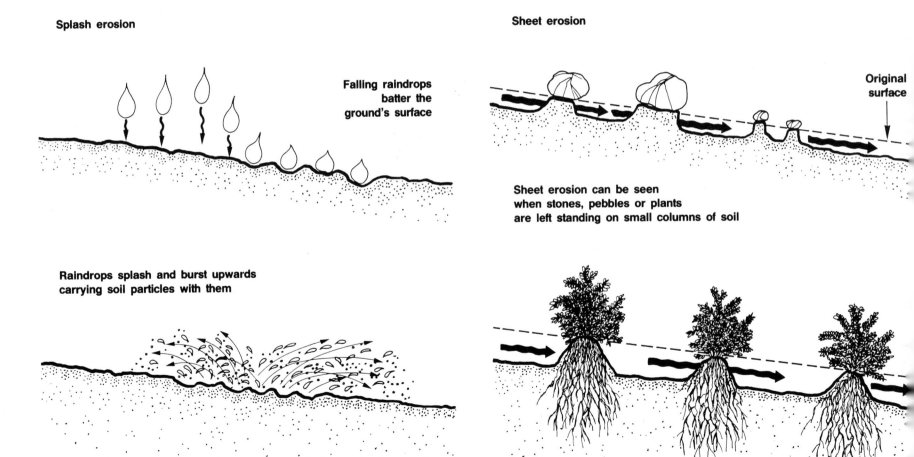

Falling raindrops batter the ground's surface

Original surface

Sheet erosion can be seen when stones, pebbles or plants are left standing on small columns of soil

Raindrops splash and burst upwards carrying soil particles with them

6. When sheet erosion remains uncontrolled, runoff cuts into the topsoil layers forming small channels. This is called **rill erosion**.

7. If heavy runoff occurs, larger flows of a water-soil mixture can gouge out deeper and deeper channels. This is called **gully erosion**.

Rill erosion

Small channels or rills caused by runoff water ...

Gully erosion

... can soon become large channels or gullies ...

8. Soil erosion essentially depends on the following factors.

(a) **The physical characteristics of the soil**, in particular its texture, structure and permeability (see **Soil, 6**).

Good soil structure

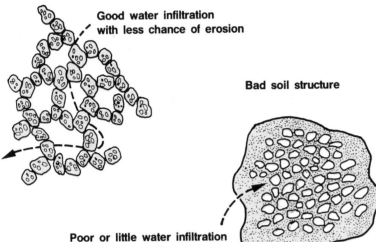

Good water infiltration with less chance of erosion

Bad soil structure

Poor or little water infiltration with greater chance of erosion

(b) **The ground slope**: as slope increases, susceptibility to erosion increases:

- flat to very gentle slope (0 to 4 percent) means little susceptibility to erosion;
- gentle slope (4 to 12 percent) means susceptibility to erosion increases rapidly;
- moderate slope (12 to 20 percent) means susceptibility is high and erosion should be controlled, especially if the slope is under cultivation;
- steep slope (over 20 percent) means that in the presence of cultivation, erosion control requires special techniques.

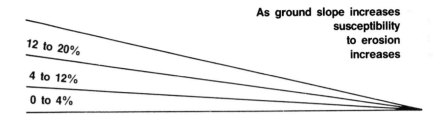

As ground slope increases susceptibility to erosion increases

12 to 20%

4 to 12%

0 to 4%

Remember: the steeper the slope, the more important and the more difficult it is to protect it against erosion.

(c) **The vegetation present**: its cover protects the soil against splash erosion. Its roots help stabilize the soil particles and increase permeability to lower soil layers. The organic matter it brings into the soil, such as humus, increases resistance to erosion and slows down runoff. It may also help soil particles to settle down.

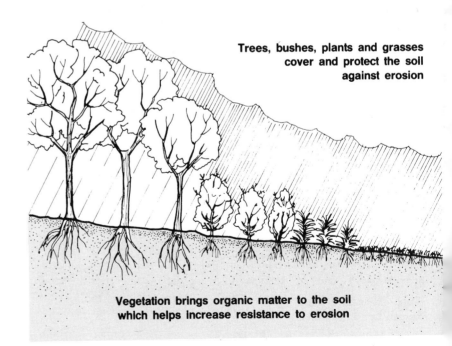

Trees, bushes, plants and grasses cover and protect the soil against erosion

Vegetation brings organic matter to the soil which helps increase resistance to erosion

9. **To control soil erosion**, you will have to use **soil conservation practices** which modify and optimize these factors. There are several ways, including:

- management of natural vegetation;
- controlled cultivation practices;
- use of physical controls.

10. You will briefly learn about these various methods in the following paragraphs. If you plan to use any of them, it is best to ask for detailed advice from the local extension agent responsible for soil conservation.

11. By using soil conservation from the beginning, you will **prevent the formation of gullies**. Prevention is much easier than dealing with gullies once they are formed. It is most important **to stabilize any incipient gully** as soon as possible to keep it from becoming longer, larger and deeper. Otherwise it might become, if not impossible, at least extremely difficult and expensive to control. Possible methods for the stabilization of gullies are described in the *FAO Conservation Guide, 13/2*, FAO watershed management field manual, Gully control, 1986.

Managing natural vegetation to conserve soil

12. By managing natural vegetation on sloping ground, it is possible to ensure that the soil has a greater resistance to erosion.

a) In **forested areas**, maintain **the soil cover** as completely as possible by managing the exploitation of the trees and protecting the forest against overgrazing and fires. Woodlands with good undergrowth, widespread root systems and good leaf cover give the best conditions.

b) In **savannah areas, control the use of fire** for the regeneration of pastures and give preference **to early fires** to ensure sufficient new growth before rains start. Avoid overgrazing, especially with sheep and goats. As far as possible, plan for **rotational grazing**.

Maintain forest cover by managing exploitation and protect against overgrazing and fires

Forested areas

Control the use of fire to ensure new growth before rains start; protect against overgrazing

Savannah areas

Cultivating soil to conserve it

13. When soil is placed under cultivation, **conservation practices** may include the following.

(a) Avoid repeating the cultivation of the same crop by **rotating crops** and keeping the soil covered as long as possible, especially at the beginning of the rainy season.

(b) Improve **the soil cover** by increasing fertilization and crop density. Plan sowing and harvesting so as to have the soil covered during the heaviest rains.

(c) Associate **several crops together** to maximize soil protection.

(d) Between two culture cycles, cover the soil with **a mulch** made, for example, from the residues of the last harvested crops.

(e) Plant **cover crops**, usually legumes and grasses. Grasses are more effective and are either used alone or mixed with legumes (see **Tables 5** and **6**). They provide **excellent forage** for farmed livestock and even for some plant-eating fish such as the Chinese grass carp.

Examples of three-year crop rotations

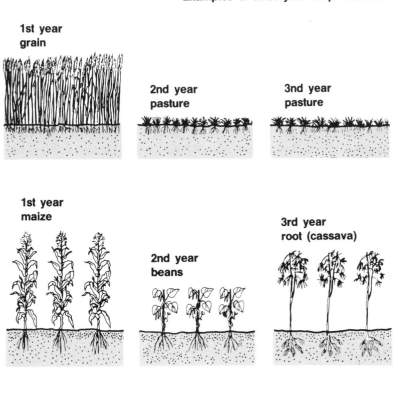

1st year grain

2nd year pasture

3rd year pasture

1st year maize

2nd year beans

3rd year root (cassava)

Plant several crops together

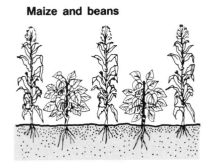

Maize and beans

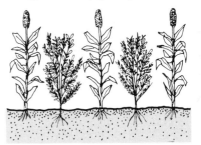

Sorghum and pigeon peas

(f) **Contour plough** the soil following contour lines (see **Topography, 16/2**). This step is most effective by itself if the soil is sufficiently pervious and if the slope does not exceed 3.5 percent.

(g) **Contour farm** following contour lines, instead of across or straight up and down the slopes. The steeper the slope, the more important it is to undertake all crop-raising activities along contour lines.

(h) Establish **perennial* vegetation strips** 3 to 8 m wide to separate 15 to 30 m wide **contour strips** under cultivation. You can use:

* **live fences** made of shrubs and trees (see **Table 7**);
* **absorbent strips** made either of grasses or of grasses mixed with legumes (see **Tables 5** and **6**); or a combination of these two types.

Note: you will learn more about this particular method for soil conservation at the end of this section.

Plough and farm sloping land on the contour ...

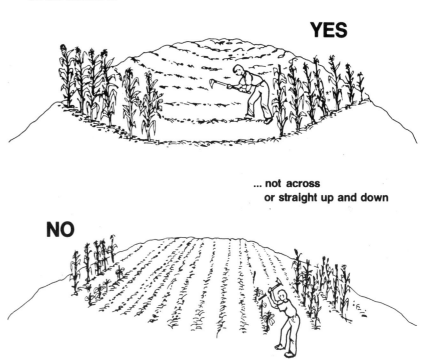

YES

... not across
or straight up and down

NO

An example of perennial vegetation planted on the contour to separate strips under cultivation

TABLE 5

Grasses useful for soil conservation

Latin name/ common name	Origin	Vegetal cycle[1]	Maximum height (m)	Climate — Air temp.	Climate — Rainfall (mm)	Resistance — Drought	Resistance — Grazing	Establishment[2]	Spacing (cm) — Pure stand	Spacing (cm) — Mixed stand	Harvesting for forage	Remarks
Cenchrus ciliaris Buffel grass	Africa, India, Indonesia	P	1.50	Warm	350-1 000	High	High	S 5 kg/ha RC 2-3 tilled	25 × 50	50 × 50	After 2-3 months, before flowering cut 10-15 cm above ground	—
Chloris gayana Rhodes grass	S.E. Africa	P	1.50	Warm to temp.	600-1 000	Medium	Low	S rows at 50 cm RC 10 cm deep	25 × 50	50 × 50	When 50 cm high, before flowering cut 10-15 cm above ground	—
Panicum maximum Guinea grass, Tanganyika grass	Trop-+subtrop. Africa	P	3.50	Warm	over 1 000	High	Low	S 4-10 kg/ha RC 10-15 cm deep	50 × 100	100 × 100	After 3-4 months, before flowering cut 10-15 cm above ground	—
Phalaris aquatica Phalaris, toowoomba, canary grass	S. Europe, Mediter-ranean	P	1.00	Temp.	430-630	Very high	Medium	S 6-10 kg/ha in rows at 50 cm	20 × 50	(no)[3]	Cut 10-15 cm above ground when first flowering shoots are seen	Useful to maintain canal/ stream banks
Pennisetum clandestinum Kikuyu grass	High tropics, E. Africa	P	1.20	Temp.	1 000-1 600	Low	High	RC + SC	30 × 30 to 50 × 50	(no)[3]	Generally grazed *in situ*	Best for *steep* slopes with fertile soil
Pennisetum purpureum Napier or elephant grass	Tropical Africa	P	4.50	Warm	over 800	Medium	High	RC + SC	30 × 50 to 60 × 60	(no)[3]	Grazing when 1-1.2 m tall maximum; cut before 2 m tall	—
Setaria sphacelata Rhodesian grass, setaria grass	Tropical Africa	P	2.00	Warm to cool	500-700	Medium	Medium	RC 10-15 cm deep S 5-10 kg/ha in rows	50 × 100	50 × 100	Cut 10-15 cm above ground before flowering	Useful in cool areas
Sorghum almum Columbus grass	Argentina	P	4.50	Warm to temp.	400-900	Medium	Low	S 5-7 kg/ha in rows at 80-100 cm	80 × 80 to 100 × 100	(no)[3]	Grazed from 35 cm high, cut 15-20 cm above ground at early flowering stage	Easily established; more nutritious when young
Tripsacum fasciculatum Guatemala or Honduras grass	S. America	P	3.50	Warm	over 1 000	Low	Nil	—	—	—	Cut 25 cm above ground	Unsuitable for grazing Tolerates acidity

[1] P = perennial
[2] S = seeds; RC = root cuttings; SC = stem cuttings
[3] (no) = possible under proper management

TABLE 6

Legumes useful for soil conservation

Latin name/ common name	Origin	Vegetal cycle[1]	Maximum height (m)	Climate		Resistance		Establishment[2]	Harvesting	Remarks
				Air temp.	Rainfall (mm)	Drought	Grazing			
Desmodium uncinatum Silverleaf desmodium, Spanish clover	C. + S. America	P	—	Warm	Over 900	Good	Medium to low	S 2-4 kg/ha in rows at 50-100 cm	Generally grown mixed with grasses N fixed = 90-160 kg/ha. Seed yield: 200-300 kg/ha	Soak seeds for 24 hours Inoculate seeds with special *Rhizobium* bacteria before sowing
Lablab purpureus Lablab, Tonga bean, lubia, batao, wal, frijol, jacinto, dolique d'Egypte	Tropics esp. Africa	A/P	—	Warm to cool	Wide range	Good	Low	S 5-7 kg/ha in rows at 100 cm	Usually grown in association with sorghum or maize. Cut above 25 cm high. Seed yield: 1.5-2 t/ha	Up to 2 000-m altitude Fast growth but short life
Macroptilium atropurpureum Siratro	C. + S. America	P	—	Warm to temp.	800-1 600	Good	Low	S 4-8 kg/ha in rows at 50-80 cm	Good results when sown with grasses Cut/graze high, at longer intervals Seed yield 200-300 kg/ha or more	Soak seeds for 24 hours Altitude: up to 1 600 m
Stylosanthes guianensis Stylo, alfalfa del Brazil, tropical lucerne	C. + S. America	P	1	Warm to temp.	Over 900	High	Low	S 1-3 kg/ha in rows at 50-100 cm	Good results when mixed with grasses Cut 15 cm high after 6-8 weeks. Seed yield 100-200 kg/ha. Rotational grazing	Soak seeds over night
Pueraria phaseoloides Puero, tropical kudzu	S.E. Asia	P	(5-6)	Warm	Over 850	Low	Low	S in rows at 100 cm SC 70-100 cm	Good compatibility with guinea and napier grasses. High nitrogen production in soil	Prefers altitude below 1 000 m. Soak seeds in 60°C water for several hours. Slow establishment

> [1] = perennial; A = annual
> [2] = seeds; SC = stem cuttings

14. These controls generally involve **extensive earth movements**, often difficult to design and costly to realize. If the above practices are not sufficient to control erosion on your ground, especially if you wish to farm a rather steep slope year after year, you should look for **specialized advice from your agricultural extension agent**.

15. There are several possible physical controls, according to local conditions of slope, soil quality and rainfall characteristics. These are:

- contour berms;
- infiltration ditches;
- contour terraces;
- gulley check dams.

(a) **Contour berms** are a series of horizontal or slightly sloping ridges constructed to intercept and temporarily store runoff. The berm foreslopes may be planted with perennial vegetation to improve their resistance to erosion. Contour berms are useful in rather dry areas with considerable runoff after heavy rains, on slopes not steeper than 20 percent.

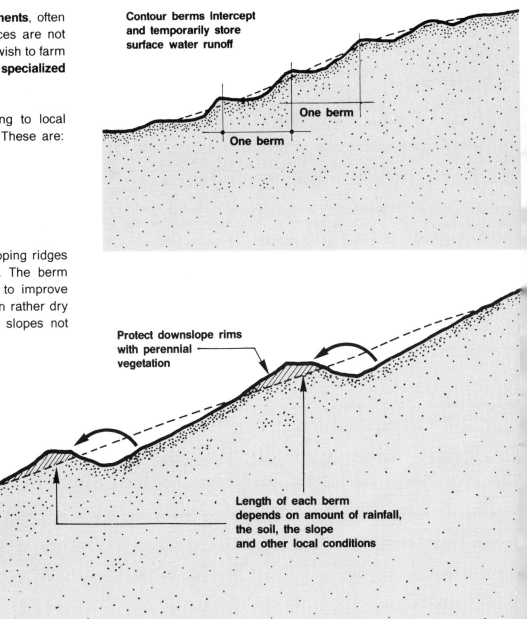

Contour berms intercept and temporarily store surface water runoff

One berm

One berm

Protect downslope rims with perennial vegetation

Length of each berm depends on amount of rainfall, the soil, the slope and other local conditions

Profile showing construction of contour berms

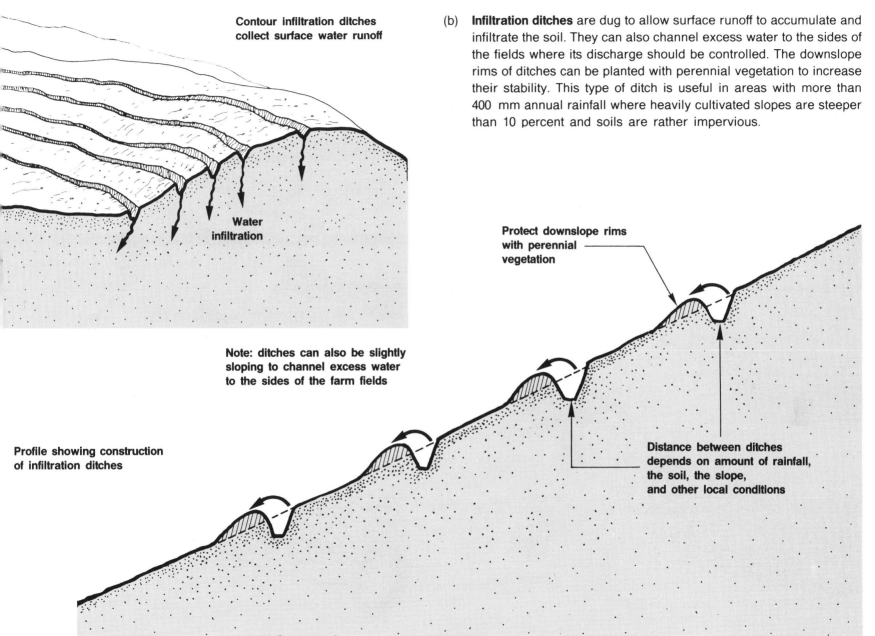

Contour infiltration ditches collect surface water runoff

Water infiltration

(b) **Infiltration ditches** are dug to allow surface runoff to accumulate and infiltrate the soil. They can also channel excess water to the sides of the fields where its discharge should be controlled. The downslope rims of ditches can be planted with perennial vegetation to increase their stability. This type of ditch is useful in areas with more than 400 mm annual rainfall where heavily cultivated slopes are steeper than 10 percent and soils are rather impervious.

Protect downslope rims with perennial vegetation

Note: ditches can also be slightly sloping to channel excess water to the sides of the farm fields

Profile showing construction of infiltration ditches

Distance between ditches depends on amount of rainfall, the soil, the slope, and other local conditions

95

(c) **Contour terraces**, providing strips of level or near-level farmland, can be built behind steep banks protected by perennial vegetation or, if sufficient stones are available, behind stone retaining walls. Contour terraces are useful on steep slopes and on those with considerable runoff after medium rains.

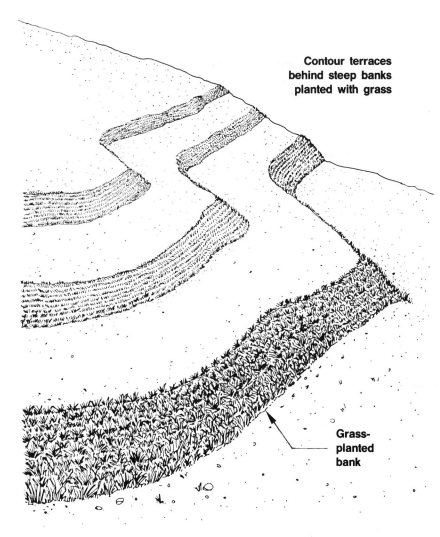

Contour terraces behind steep banks planted with grass

Grass-planted bank

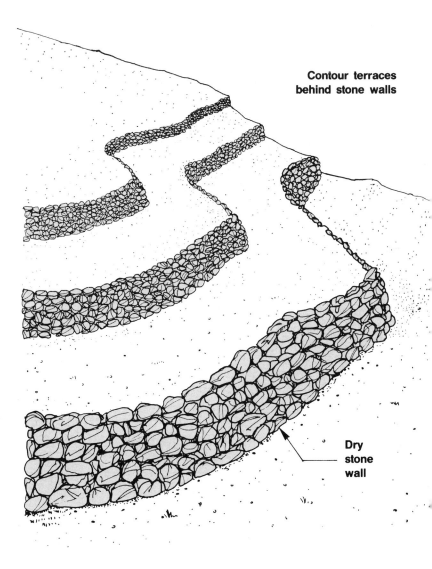

Contour terraces behind stone walls

Dry stone wall

**A check dam
of stone**

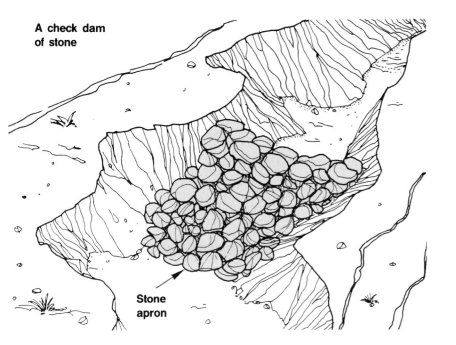

Stone
apron

**A check dam
of woven branches
and wooden poles**

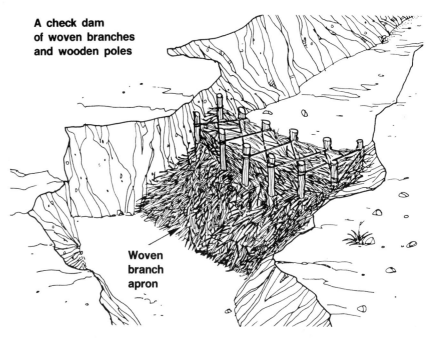

Woven
branch
apron

d) **Gulley check dams** are small dams built across gullies or small stream courses. They hold up runoff, limit downstream erosion and are useful in dry areas with heavy runoff and medium to steep slopes. A gully check dam may be made using a variety of locally available materials such as stone or woven tree branches lashed in place between two rows of wooden poles. It is very important that the top of the check dam be lower than the crest of the gully so that overflowing water remains in the gully.

Note: remove 15 to 30 cm of soil from the gully bottom and sides before placing the check dam and apron.

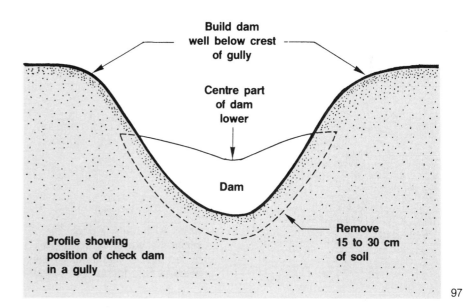

Build dam
well below crest
of gully

Centre part
of dam
lower

Dam

Remove
15 to 30 cm
of soil

Profile showing
position of check dam
in a gully

16. If you plan to produce cash crops on land with **a gentle to moderate slope** (4 to 20 percent), you may use a simple method to control soil erosion, namely to encourage **the progressive build-up of nearly horizontal terraces by controlling sheet erosion itself**.

17. Narrow contour strips are planted with **perennial vegetation** fairly closely together on the slope to be protected. Generally, the steeper the slope, the more vegetation strips you will need. If necessary, **a dead filter**, made of slightly buried branches and crop residues in front of each vegetation strip, will help to stabilize the vegetation and to accelerate terrace formation. The vegetation slows runoff, resulting in the deposit of transported soil particles, first within the vegetable strips and then further up the slope. The roots of the vegetation become buried deeper and deeper as the downslope bank of the terraces builds up.

Note: as the soil is deposited within the grass, the roots are buried deeper and deeper, which helps to hold the terrace banks

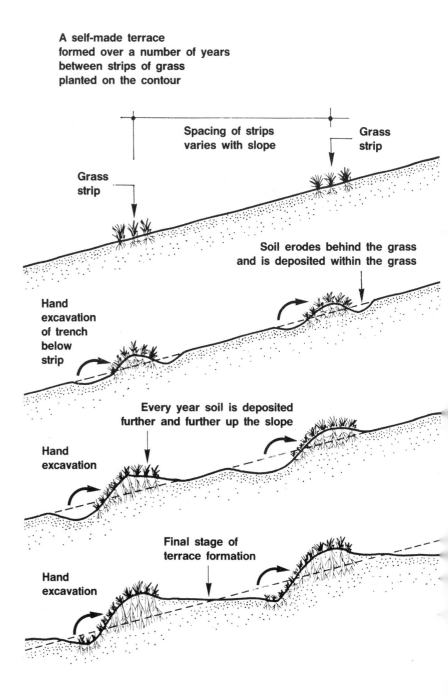

A self-made terrace formed over a number of years between strips of grass planted on the contour

Spacing of strips varies with slope

Grass strip

Grass strip

Soil erodes behind the grass and is deposited within the grass

Hand excavation of trench below strip

Every year soil is deposited further and further up the slope

Hand excavation

Final stage of terrace formation

Hand excavation

Other ways to begin
self-made terraces
on the contour

Unploughed strip

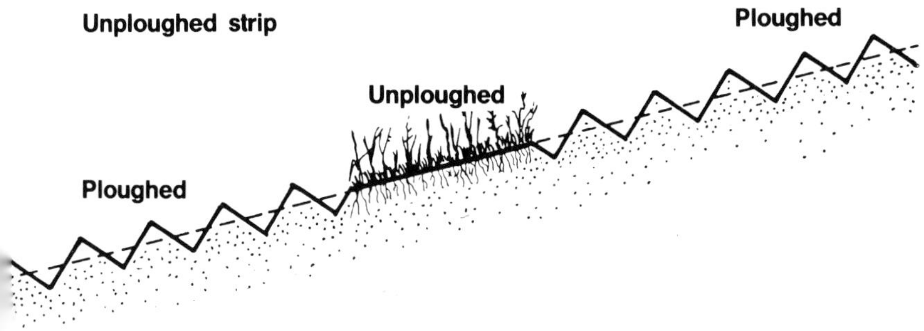

Ploughed

Unploughed

Ploughed

Dig a ditch
and throw earth uphill
(sometimes called *fanya ju*)

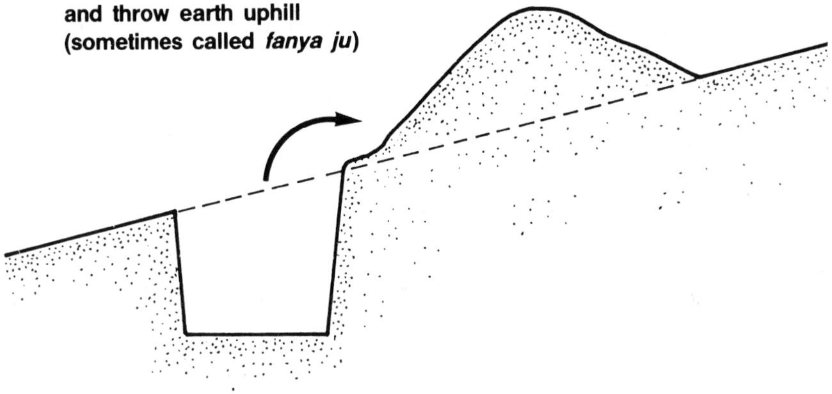

Stems, stalks,
harvest wastes

Trash strip

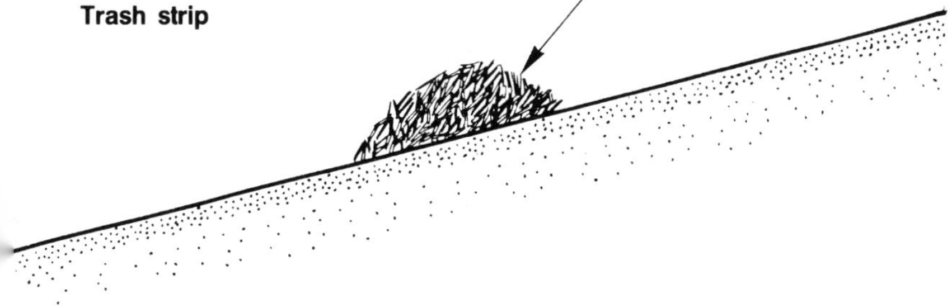

Note: all of the systems shown on this page
will form self-made terraces over the years,
very much like the terraces formed by grass
strips shown on page 98

Dry stone
barrier

Grass strip
on small ridge

Before building,
dig a shallow trench
to keep stone barrier
from sliding downhill

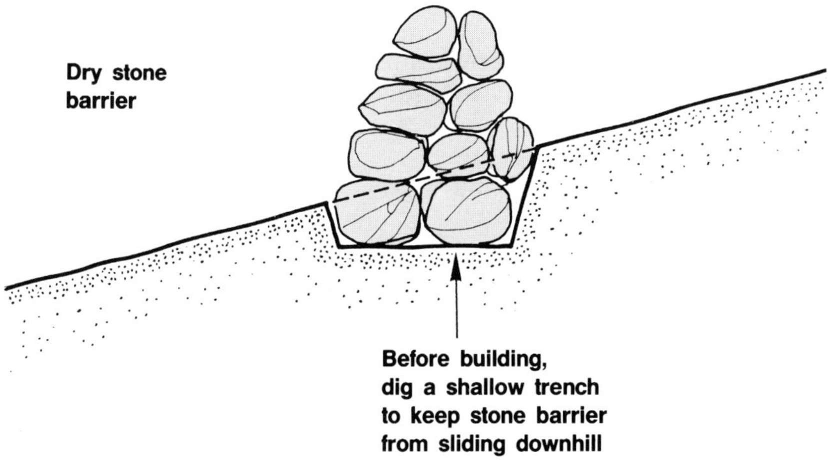

18. **Perennial vegetation strips** can be made following the contour lines in several ways.

(a) **Spontaneous vegetation** can become established if you stop farming narrow strips of land.

(b) **Forage grasses** in particular **and legumes** can be planted to provide food for livestock, either under controlled grazing or as cut forage, fed fresh or dried (see **Tables 5** and **6** above). Similarly you could use **lemon grass, sugar cane or sisal** if a market exists for such crops.

(c) One row of **shrubs and trees could be added to this forage vegetation** to produce timber, firewood, fruit or additional fodder for your livestock. Suggested species, which should be selected according to local conditions of climate and soil, include:

- shrub and tree legumes for fodder (see **Table 7**);
- other trees for diversified production (see **Table 8**);
- fruit trees can be planted for the production of mangoes, avocados, pawpaws, bananas and various citrus fruits according to local demand and market potential. If possible use genetically improved varieties.

(d) **A live fence** can be made of (more densely planted) shrubs and trees (see above for suitable species).

19. Perennial vegetation strips should be established in several phases.

(a) **Determine and mark the contour lines** at regular intervals, normally spaced to provide a farmed strip 15 to 30 m wide (see **Topography, 16/2**).

(b) **Prepare the contour strips** just before the rains start.

(c) **Establish the vegetation** after some good rains have fallen, preferably by planting either seedlings or cuttings to cover the soil as soon as possible.

(d) **Seedlings** are produced in separate nursery areas and are transplanted to the contour strips after a few weeks. The nursery area should have well-prepared seed-beds. Seeds are sown in rows at about 1.5 to 2 cm deep and covered with well-pressed soil. Seeds can also be planted in small plastic bags containing soil, which help to retain moisture and control weeds.

(e) **Cuttings** should be healthy, about six months old and with at least two nodes. Small cuttings are planted at a 40° angle with one node in the soil. Long cuttings are buried in the soil about 10 to 15 cm deep.

(f) **Protect** all seedlings and cuttings **well from grazing** until the vegetation is well established. For shrubs and trees, this might take as long as three to five years.

(g) **Keep weeds under control**, especially around the young plants.

(h) **Establish the dead filter** just uphill of the vegetation by slightly burying wooden branches, crop residues or stipes of banana leaves in the soil.

20. The vegetation strip should be **regularly maintained**.

(a) **Repair and strengthen any breaches** through the strip by planting additional vegetation in the fragile zones.

(b) As the terrace builds up, place **new dead filters** uphill of the buried ones. Repair these filters as needed.

(c) Watch for **localized erosion** either of the lower lip of the terraces or of their steeper banks; immediately repair, adding vegetation.

Dealing with steep slopes

21. On slopes steeper than 15 to 20 percent, it may not be sufficient to establish perennial vegetation strips to control soil erosion adequately. The area of land suitable for farming will also be much reduced with this method, owing to the reduced spacing between vegetation strips.

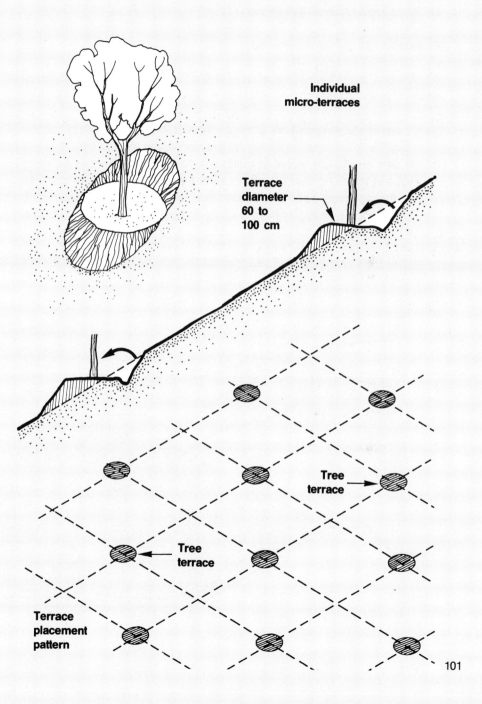

Individual micro-terraces

Terrace diameter 60 to 100 cm

Tree terrace

Tree terrace

Terrace placement pattern

22. You should therefore consider the following guidelines:

a) Avoid farming cash crops.
b) Limit farming to **perennial crops**.
c) Select or encourage **shrubs and trees** that protect the soil. **A natural forest**, if well managed and protected from fire, may be the best solution.
d) If you choose **to plant shrubs and trees**, it is best to build **individual micro-terraces** rather than using more expensive and more elaborate techniques. Protect these terraces from erosion by planting grasses and legumes at their raised edges. Protect your plantations from grazing, fires and uncontrolled cutting.

42 Wind protection in fish farms

Protecting fish ponds from wind

1. When a strong wind blows over a fish pond, it affects the environment in several ways.

(a) **It increases evaporation** at the surface of the pond, especially if the wind is a dry one, resulting in greater water losses.

(b) **It sets and keeps water in motion**, inducing surface water currents toward the pond dike facing the wind and returning deeper currents in the opposite direction. These currents help the transfer of heat and dissolved oxygen from surface to deeper water (see Sections 24 and 25).

(c) **It may generate waves**, which greatly accelerate the oxygenation of surface water (see Section 25), although wave action against the downwind dike may damage the latter through erosion.

(d) **If relatively cold**, it may delay the warming of small ponds intended for fish breeding and nursing early in the warmer season.

2. Although wind has definite advantages for fish farming, mainly keeping the ponds well mixed and oxygenated, there are **particular situations** in which you may wish to protect at least part of your fish farm

- to **warm up early** small breeding ponds and nursery ponds when relatively cold winds are prevalent; and
- to **reduce wind velocity** and the size of generated waves over larger ponds, for easier control of dike erosion.

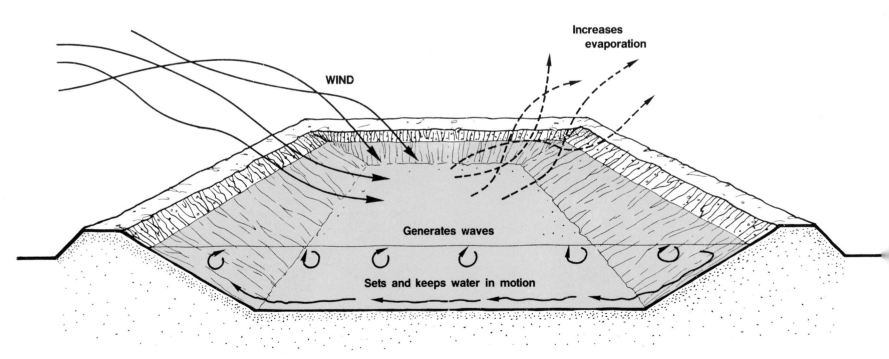

WIND

Increases evaporation

Generates waves

Sets and keeps water in motion

A simple way to provide wind protection is to build **a simple screen**, using for example canvas, plastic, or woven split bamboo stretched between vertical wooden poles. The screen has the advantage of being easily set up when and where necessary. However, it only provides protection over a very limited distance behind the screen, perhaps three to six times its height, depending on the conditions.

If you require more permanent protection over a greater surface area, it is usually better to establish **a live wind-break** made of perennial vegetation and designed to reduce wind velocity immediately above ground level. This kind of barrier also has some additional advantages.

) It provides the opportunity for **increasing the range of products** and increasing the profits of the fish farm through the production of wood for construction and fuel, animal fodder, fruit, tannin, fibres, etc.

) It **provides shade** not only for livestock but also for carrying out certain management tasks such as the sorting of live fish and the maintenance of equipment.

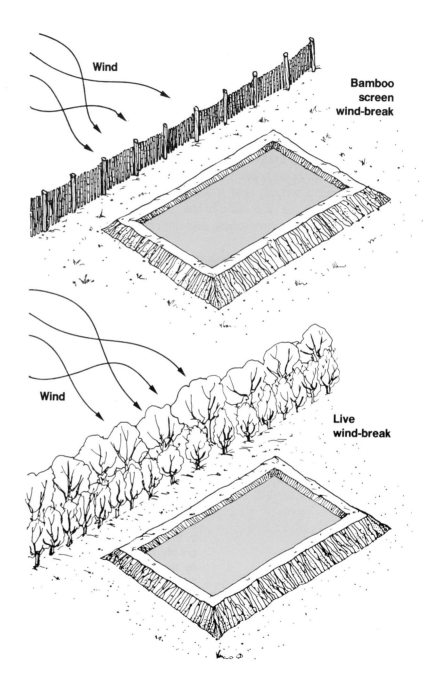

Wind

Bamboo screen wind-break

Wind

Live wind-break

103

5. An efficient wind-break should have certain characteristics. It should be:

- **semi-permeable** especially in its lower part, about 40 percent of its total area consisting of small openings regularly distributed (a lower permeability reduces the wind velocity more; turbulence, however, reduces the length of the protected area as only the density of the first row directly exposed to the wind modifies its velocity; an open breach in a wind-break will, on the contrary, increase wind velocity and it should be avoided);
- **relatively thin**: as the wind-break becomes thicker, its permeability to wind decreases and therefore its efficiency decreases;

- **rectangular in cross-section** to minimize turbulence in the protected zone;
- **as high as possible**, because the higher the wind-break, the longer the protected zone behind it;
- **long enough** to avoid side turbulence in the protected zone (the length of the wind-break should be equal to at least 12 times its maximum height);
- **perpendicular to the wind direction**.

Note: carefully check on the prevalent direction of the wind against which you seek protection. If this direction is too variable, you may either increase the length of the wind-break or establish several ones in different directions.

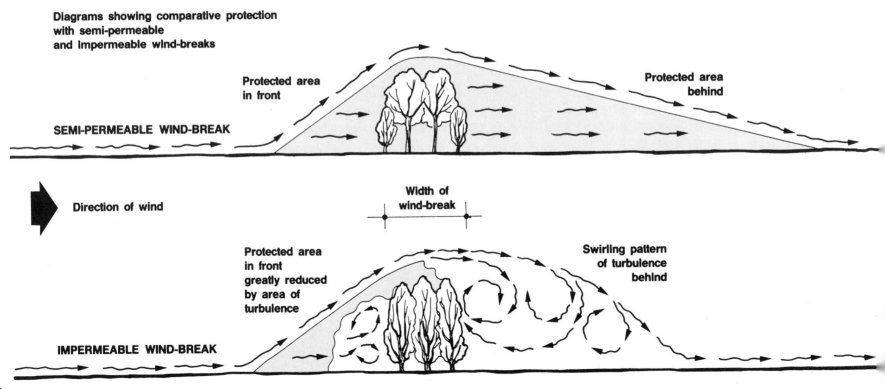

Diagrams showing comparative protection with semi-permeable and impermeable wind-breaks

Protected area in front

Protected area behind

SEMI-PERMEABLE WIND-BREAK

Direction of wind

Width of wind-break

Protected area in front greatly reduced by area of turbulence

Swirling pattern of turbulence behind

IMPERMEABLE WIND-BREAK

Designing a live wind-break

When designing a live wind-break, pay attention to the following points:

- location of the wind-break;
- use of tall trees;
- use of smaller trees, shrubs or tall grasses.

Plant the wind-break at least 3 m outside the centre line of pond dikes and even further if the trees are likely to have long horizontal roots.

Use at least one continuous row of tall trees. If necessary, add one or more rows of smaller trees, shrubs or even tall grasses to complete the semi-permeable screen in its lower part.

Protection characteristics possible when designing a wind-break

(c) The protected zone will extend slightly in front of the wind-break and much further behind, with the wind velocity being reduced in proportion to the maximum height (H in m) of the wind-break:

- by 20 percent from about 1 × H in front to about 10 to 15 × H behind;
- by 50 percent from about 0.5 × H in front to about 1.5 to 2.5 × H behind.

Example

The maximum height of the wind-break is 10 m. The protected zone will extend at the most from about 10 m in front to 150 m behind the wind-break.

(d) Preferably, **plant several rows of trees**:

- to avoid breaches forming in the wind-break if, for example, certain trees die from disease;
- to enable **the exploitation of mature trees** while leaving at least one row offering protection; and
- to improve **the survival of young trees**, which resist strong winds better when planted on a wider strip.

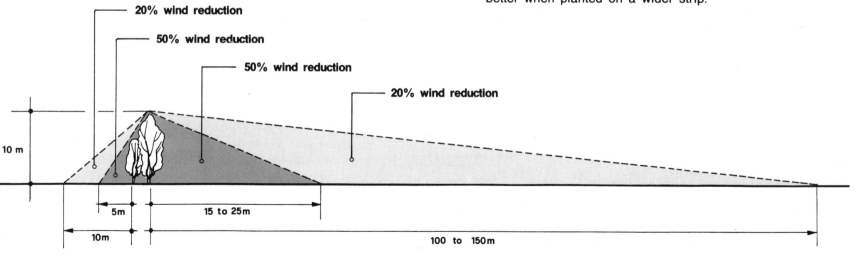

Direction of wind

20% wind reduction
50% wind reduction
50% wind reduction
20% wind reduction

10 m

5m
15 to 25m
10m
100 to 150m

105

(e) To protect large areas, you may require **a series of wind-breaks** parallel to each other. The distance between each wind-break should be **about 15 times** their maximum height.

(f) Select **tree species well adapted** to local conditions of climate and soil. Seek advice from your forestry or agriculture extension agent.

(g) **Diversify the benefits** of your wind-break as much as possible, according to your own needs and those of your community. Adapt the **volume of production** of each particular product to local demand.

(h) In addition to these last two considerations, **give preference to tree species** with the following characteristics:

- sufficient **height** for the area to be protected;
- **foliage homogeneous** from top to bottom of crown, and evergreen foliage permeable to wind;
- **fast growth** in height;
- reduced **crown volume**;
- strong **taproot** and limited horizontal root system; and
- easy **regeneration**, either naturally by seeds or by **coppicing***.

At least four rows of trees for best efficiency

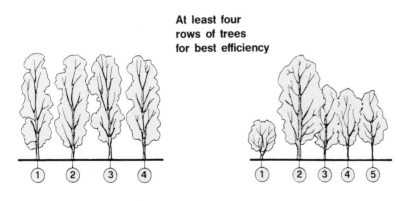

Note: best efficiency has been claimed for wind-breaks made of at least **four rows of trees** occupying a strip of land 10 to 12 m wide.

7. Very few tree species have all these desirable characteristics, and therefore it is usually necessary **to mix different species** to ensure that the wind-break has the desired efficiency and structure.

8. In most cases, you will require **tall trees** to increase the size of the protected area and to produce various wood products; depending on the local conditions you may use:

- leguminous trees such as *Acacia albida*, *A. auricoliformis*, *Albizzia lebbek* or *Tamarindus indica* (see **Table 7**), *Cassia siamea*, *Prosopis chilensis*;
- some fast-growing but relatively short-living trees such as *Azidarachta indica* (neem), *Eucalyptus camaldulensis* (needs at least three dry months per year), or *Casuarina equisetifolia* (see **Table 8**);
- some slower-growing but long-living trees such as *Parkia biglobosa* (see **Table 8**) and *Cupressus* spp.

9. You will also need **small trees and shrubs** to complete the wind-break below the crown of the tall trees. These species will mainly produce firewood, animal fodder and fruit. Depending on the local conditions, you may use some small legumes (see **Table 7**) or fruit-bearing trees (*Anacardium occidentale*, cashew; *Morus* spp., mulberry).

10. At the lowest level of the wind-break, you could also plant:

- perennial **forage grasses** such as *Andropogon gayanus* (Gamba grass for 600 to 1 100 mm annual rainfall with five to six dry months a year), *Chloris gayana* (Rhodes grass for 600 1 200 mm annual rainfall with four to six dry months a year), others (see **Table 5**);
- **thorny shrubs** such as *Acacia* (*A. mellifera*, *A. senegal*) protect the wind-break from being damaged by wandering animals;
- **shrubs unpalatable to livestock** such as *Euphorbia balsamifera* (dry soils) or *E. tirucalli* (humid soils).

Leguminous shrubs and trees useful for soil conservation and wind-breaks

Latin name/ common name	Origin	Vegetal cycle[1]	Maximum height (m)	Climate		Resistance		Establishment[2]	Spacing (cm)	Harvesting for forage	Remarks
				Air temp.	Rainfall (mm)	Drought	Grazing				
Acacia albida Albida, kad, haraz, winter thorn	Tropical Africa	D	30	Warm	300-800	High	Low	SD 2.5-4 months	400/ha cleared to 10 m × 10 m	Produces soft wood (10-15 years) and animal food	Soak seeds in 80°C water, cool for 24 hrs; greatly increases soil fertility
Acacia saligna Saligna	W. Australia	5-10 yrs	4-5 (coppice)	Warm	200-1 000	High	Low	SD 3-4 months	100 × 200	Harvest above 50 cm high, at long intervals; produces firewood, fencing, gum, animal food	Soak seeds in 80°C water, cool for 24 hrs; ability to coppice (suckers)
Albizia lebbek Albizia, siris, mataration, tibit tree	Africa, Asia, N. Australia	D	20 (coppice)	Warm to cool	Over 600	Good	Low	SD 2.5-3 months	500 × 500 to 800 × 800	Produces good timber (10-15 years), firewood and fodder	Soak seeds in 80°C water, cool for 24 hrs; ability to coppice
Atriplex nummularia Oldman saltbush	Australia	P	3	Warm to temp	Over 150	High	Low	SD 15-20 cm SC transplants	—	Grazing when 1.5 m high	Soak seeds in water for 2-3 days, changing water 2-3 times daily; to stabilize sandy soils
Cajanus cajan Pigeon pea, guando, dahl, catjang, pois d'angole	Egypt, E. Africa, India	P/D (up to 5 yrs)	1-3	Warm to temp	600-1 000	Good	Low	S 1-4 kg/ha in rows at 100 cm	30 × 100 to 50 × 100	Rotational grazing; hand harvest at 20-30 cm in height; seed yield 1-1.2 t/ha	—
Leucaena leucocephala Leucaena, ipil-ipil, aroma blanco	Mexico	P/D	Shrub tree (20 m)	Warm	Over 500	Good	Low	ST < 10 cm diam. SD 3 months S 2-10 kg/ha in rows at 100-150 cm	Variable	Harvest at 50 cm in height at 10-12-week intervals before flowering; seed yield 1-1.5 t/ha; rotational grazing	May produce hardwood for charcoal; soak seed 2 min in 80°C water; wash with cold water and dry; inoculate with special *Rhizobium* bacteria
Prosopis glandulosa Prosopis, mesquite	S.W. USA, Mexico	D	Shrub tree (4-8 m)	Warm	150-600	High	Low	SD 3-3.5 months	5 m diam. or more	Good source of firewood, timber, charcoal, human/animal food	Good for sandy soils; soak seeds in 80°C water, cool for 24 hrs; aggressive, invades
Sesbania grandiflora Sesbania, gallito, cresta de gallo	Asia	P	Shrub tree (10 m)	Warm	Over 1 000	Low	Low	S 10 kg/ha in rows at 100-200 cm SD 15.2 months	—	Harvest 3-4 months after planting; produces firewood, paper, gum, human/animal food	Grows very rapidly; top the seedlings off after 1-1.5 months transplant
Tamarindus indica Tamarind, tamarinier	Ethiopia, C. Africa	P	15	Warm	—	Good	Low	—	—	Produces wood, good fodder and fruit	Slow growth; drought resistant

= perennial; D = deciduous
= seeds; SC = stem cuttings; SD = seedlings; ST = stumps

TABLE 8

Non-leguminous shrubs and trees useful for soil conservation and wind-breaks

Latin name/ common name	Origin	Maximum height (m)	Climate[1]	Diversified production							Remarks
				Timber	Poles	Posts	Firewood	Charcoal	Fruit	Miscellaneous	
Anona reticulata Bullock heart, corazón, coeur-de-boeuf	Tropical C. America	8-10	Warm, humid	—	—	—	—	—	●	Fibres, insecticide, tannin, dye	—
Bambusa vulgaris Yellow, feathery bamboo	Pacific Islands	15	P > 1 000 mm Alt. < 1 500 m	—	●	—	—	—	—	—	Plant 4 × 4 m; clear at 8 × 8 m; selective harvest, leaving at least 8 culms
Butyrospermum parkii Shea-butter tree, karité	Africa	9-12	P 700-1 000 mm	—	—	●	●	●	●	Shea-butter	Resists 4-5 dry months; slow growth
Casuarina equisetifolia Filao	S.E. Asia	High	Low altitudes	●	—	●	●	●	—	—	Excellent for sandy soils; plant 5- 8- month seedlings at 3 × 3 m
Colubrina arborescens Wild ebony, abeyuelo, bois pelé	C. + S. America	20	Wide range	●	—	●	—	—	—	—	Fast growth in humid climate
Grevillea robusta Silk oak, roble de seda	Australia	25	Drought resistant	—	—	—	—	—	—	Wood for furniture and interiors	Fast growth; easy propagation
Oxytenanthera abyssinica Bamboo	Africa	10	Dry 900-2 000 m	—	●	—	—	—	—	—	Plant 4 × 4 m; clear at 8 × 8 m; selective harvest, 25% of culms
Parkia biglobosa African locust bean, néré, caroubier africain	Africa	10-20	P 600-900 mm	—	—	—	●	—	●	—	—
Schaefferia frutescens Floride-boxwood, jiba, limoncillo	C. + S. America	10	Drought resistant	—	—	—	—	—	—	Wood for furniture	Evergreen shrub/tree; fast growth
Simaruba glauca Simaruba, marupa, aceituno	C. + S. America	20-30	—	●	—	—	—	—	—	Food (seeds), oil	Fast growth

[1] P = annual rainfall

Various kinds of wind-breaks

1. Various kinds of wind-breaks can be planted, usually made of **two to six rows**, determined not only by the kind of additional benefits expected but also by the land available. Common examples of wind-breaks are the following:

Two-row wind-break

Row 1
Shorter tree or shrub
Row 2
Tall tree, fast growth

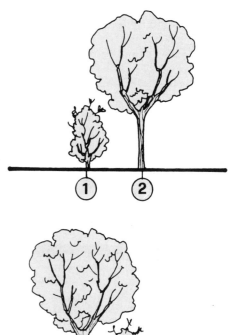

Wind

a) **Two-row wind-breaks**: one fast-growing tree planted with a smaller tree. The smaller tree can be planted either in front of or behind the larger one, according to the resistance to the wind of both kinds of trees.

Two-row wind-break

Row 1
Tall tree, fast growth
Row 2
Shorter tree or bush

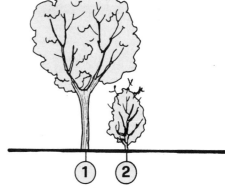

Three-row wind-breaks: two rows of tall trees, one of which may be fast-growing but relatively short-living (e.g. *Eucalyptus* spp. or *Azidarachta indica*) and the other one may be slow-growing but long-living (e.g. *Acacia albida* or *Tamarindus indica*); and one row of smaller vegetation producing firewood, fodder, fruit, etc. Major wood production is expected from the exploitation and **even removal of the middle row** when the wind-break becomes too dense

Three-row wind-break

Row 1
Shorter tree or shrub
Row 2
Tall tree, fast growth, short life
Row 3
Tall tree, slow growth, long life

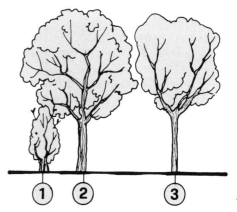

(c) **Four-row wind-breaks**: especially useful for additional benefits such as fodder, poles and firewood. The four rows may be planted with one tree species such as a leguminous tree. In dry locations, consider *Cassia siamea* and in wet locations, *Dalbergia sissoo*. Plant these trees in alternate rows, with about 3 m between trees.

(d) **Five-row wind-breaks**: from front to back row, the following species can be planted: *Cajanus* sp., *Casuarina equisetifolia*, *Acacia nilotica* (or two inside rows of *Cassia siamea*).

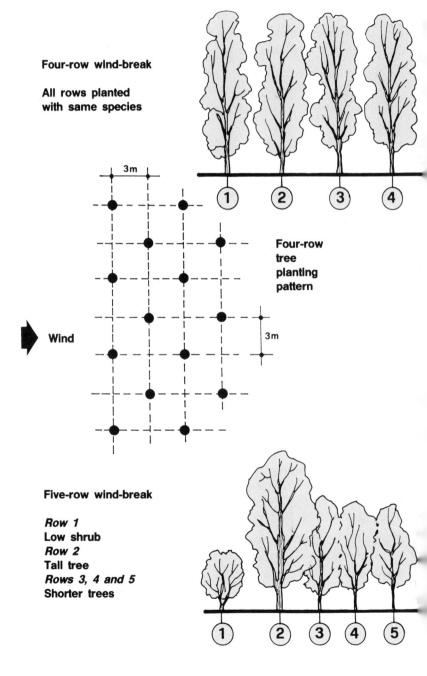

Four-row wind-break

All rows planted with same species

Four-row tree planting pattern

Wind

3m

3m

Five-row wind-break

Row 1
Low shrub
Row 2
Tall tree
Rows 3, 4 and 5
Shorter trees

2. **To establish your wind-break**, proceed in stages:

a) Obtain or produce **vigorous plants of uniform size** for the tree species you have selected.

b) **Prepare the strip of land well** for planting by loosening the soil to a depth of 0.6 to 0.8 m and by digging deep holes. Fertilize if the soil is infertile.

c) After the rains have started, **plant the young trees**.

d) **Protect them** from being damaged by grazing animals and fire.

e) **Replace** all dead plants as soon as possible.

f) **Water the plants** if necessary, at least until they have developed a deeper root system.

g) **Eliminate the weeds and loosen the soil** around the plants regularly.

3. Later, once the wind-break is established and growing fast, it should be maintained regularly, which may also provide wood products.

a) **Remove and if necessary replace trees** which are dead, diseased or much slower-growing.

b) If the permeability of the screen becomes too low or if the tree growth starts declining, **reduce tree density**.

c) **Pollarding*** may also become necessary.

d) Continue **to protect the plants** against overgrazing, fire and uncontrolled cutting.

e) If necessary, improve the wind-break efficiency by adding one or more new rows of vegetation.

4. When the trees of a wind-break reach maturity, their ability to reduce wind velocity decreases greatly. **Exploiting and renewing** them becomes necessary. The most common practice is **to first exploit one half of the wind-break** by cutting one or two rows of trees at ground level. This vegetation is then renewed either by **coppicing** or by planting, and the other half of the wind-break is cut down a few years later. This system is particularly easy to manage with a four-row plantation, but a similar scheme can be adopted for other types of wind-break.

Examples of wind-break exploitation and renewal

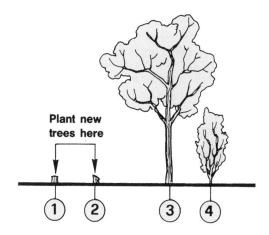

Cut two rows of trees, leaving two rows as wind-break

Plant new trees here

① ② ③ ④

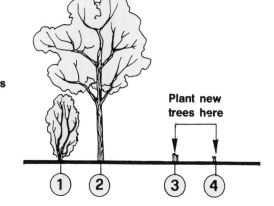

A few years later, when the new trees have grown, cut the other two rows

Plant new trees here

① ② ③ ④

43 How to protect dikes against wave action

1. In ponds that are relatively large or poorly protected from strong winds, the lapping of the waves against the upper part of the dikes may cause extensive damage. Such erosion is particularly active:

- in newly constructed ponds; and
- on the wet sides of the pond dikes facing the wind.

2. **To protect dikes** against wave action, you can use one of the following methods:

(a) Around the normal water level and over the area to be protected, lay **one or more rows 1 to 1.5 m wide of wood or bamboo poles**. Place them close to each other. Anchor each row well using wooden stakes, for example, so that the poles do not float away.

(b) You can also use **split bamboo, rush or fibre matting** laid in mats 1.5 to 2 m wide and anchored into the dike. They will not last so long, but are cheap and can be easily moved or replaced. Plastic sheeting or large leaves from palm or banana, for example, can also be used for temporary protection.

(c) Build **a fence of wooden or bamboo poles** driven side by side into the dike slope about 0.5 m from the water's edge when the water is at its normal level. To reinforce this fence, you can lace the tops of the poles together with lianas or ropes.

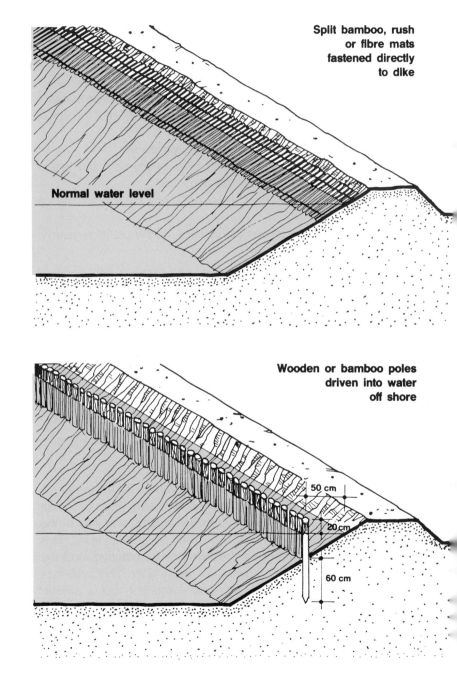

Split bamboo, rush or fibre mats fastened directly to dike

Normal water level

Wooden or bamboo poles driven into water off shore

50 cm

20 cm

60 cm

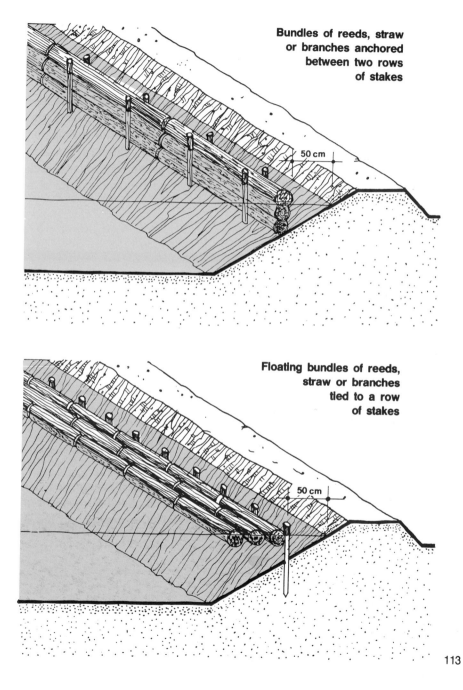

Bundles of reeds, straw
or branches anchored
between two rows
of stakes

50 cm

d) A similar fence can be built with 30-cm-thick **bundles** of reeds, straw or branches stacked and tied end to end **between two rows** of stakes.

Floating bundles of reeds,
straw or branches
tied to a row
of stakes

50 cm

Floating bundles of straw or branches may be assembled in a continuous line just in front of the dike. Make sure they extend at least 30 cm below the water level.

Rocks or stones
set into sides
of dike

(f) More permanent protection can be made **if rocks or stones** are easily available. Use them in a layer 25 cm thick spread over a strip 1.5 to 2 m wide and extending on either side of the anticipated normal water level. Coconut shells and similar materials can also be used. In extreme cases such as for large barrage ponds, **rock rip-rap**, preferably placed by hand, might be required on top of a 25-cm-thick gravel foundation.

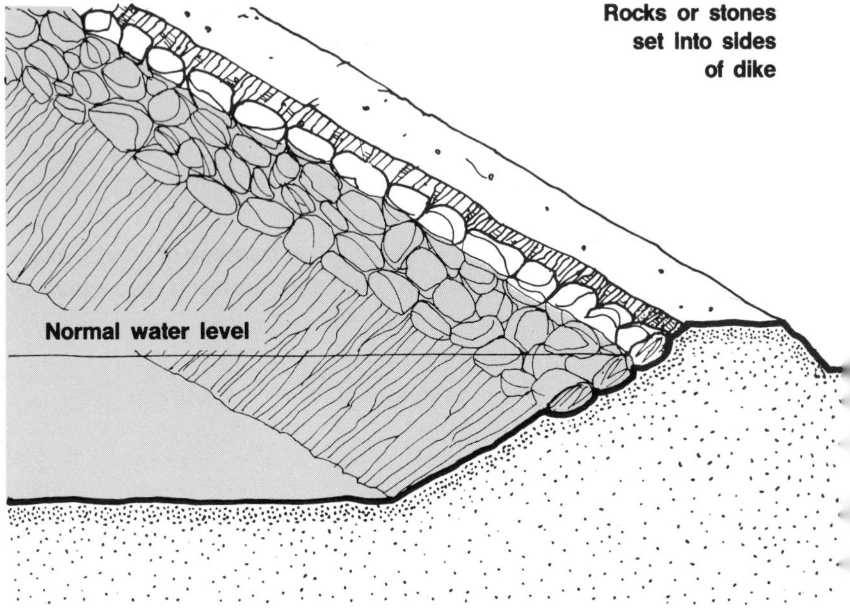

Normal water level

Floating logs tied
end to end
and anchored
off shore

(g) **A row of end-to-end floating logs**, up to 25 cm in diameter, may be anchored about 2 m offshore. This structure is called **a log boom**. If the pond water level is expected to fluctuate, remember to leave some slack in the anchor lines.

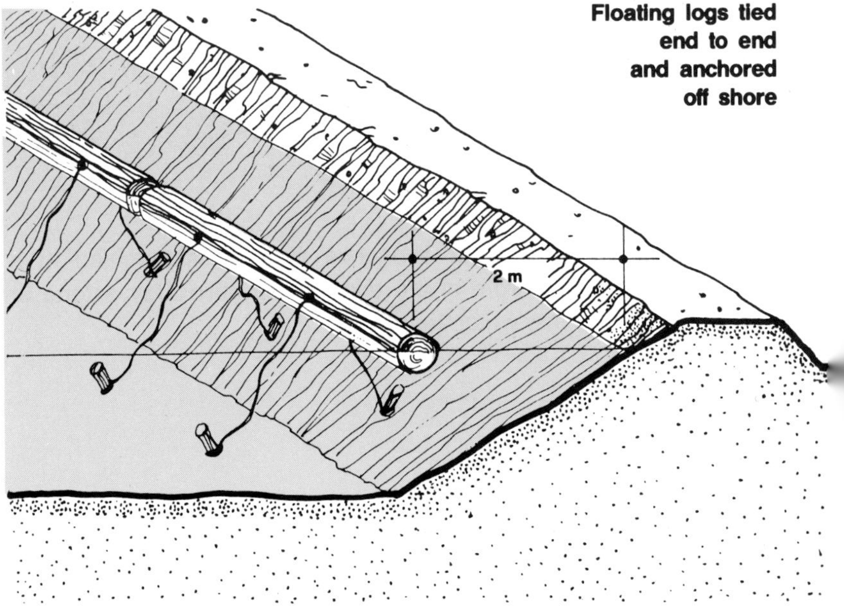

2 m

One of the best and longest lasting methods of wave action control may be obtained from **dense aquatic vegetation**. Usually a strip 2 m wide of strong vertical aquatic plants, reeds for example, is planted in front of the dike to be protected. If you plan to use this method, **you should design the dike**:

- either with **a very gentle slope** from 1:4 to 1:8 starting 0.3 m below the normal water level;
- or **with a horizontal berm** about 0.5 m below normal water level and 3 m wide.

To protect young aquatic plants from waves until they are grown, use bundles of reeds, straw or branches anchored in front of the horizontal berm as shown on page 113. When the strip of vegetation is fully grown, you can remove this protection.

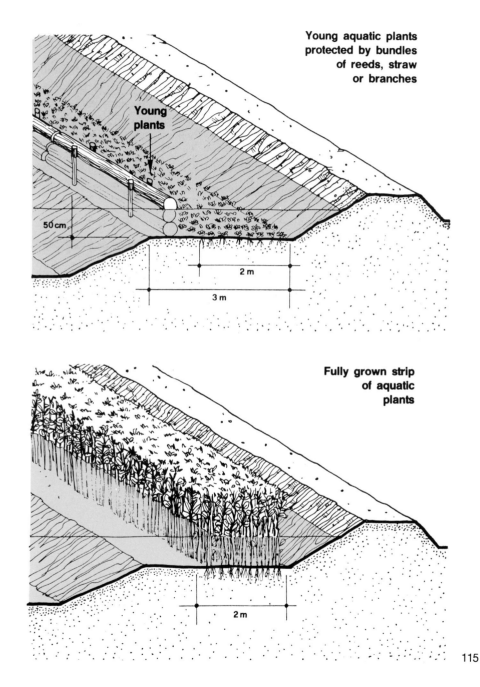

Young aquatic plants protected by bundles of reeds, straw or branches

Young plants

50 cm

2 m

3 m

Fully grown strip of aquatic plants

2 m

44 How to protect dikes and canals against erosion

Protecting your dikes

1.   You have already learned earlier that you should protect **newly built dikes** against erosion from rainfall, covering them as soon as possible with a thick grass cover (see Section 69, **Pond construction, 20/1**). You should maintain this grass cover well, cutting it regularly and fertilizing it if necessary.

2.   You have also learned how to protect your fish ponds against **wind** (see Section 42) and the resulting **wave action** (see Section 43).

3.   Dikes should also be protected against damage caused by **heavy livestock traffic**, as you will learn later (see Section 45).

4.   Dike erosion can also take place **at the pond inlet** under the action of the inflowing water. Protection may be given either:

- **directly** to the part of the dike located under the inflowing water (see Section 20, **Pond construction, 20/2**); or
- **indirectly** through some of the water filtering devices described earlier (see Section 29), which absorb the force of the inflowing water.

Protecting your water canals

5.   Water feeder, drainage and diversion canals are subject to erosio particularly if **the water current** exceeds the maximum permissib velocity (see Section 82, **Pond construction, 20/2**). Sheet erosion durin rainfall may also take place along the canal side slopes if they are n properly protected.

6.   There are different ways of providing additional protection to wat canals. You can do this in one of the following simple ways.

(a)   First take any steps possible **to reduce water flow** in affect sections (e.g. by draining ponds carefully) **or to reduce she erosion** (e.g. by diverting water away from the side slopes w small trenches).

(b)   You may connect the trenches with **gravel filter channels**, if y need to conduct the water safely down the canal side. These a strips of gravel 30 to 40 cm wide and 20 to 30 cm deep, which r diagonally down the side of the slope.

(c)   Establish and maintain a strong short **grass cover** on the ca sloping sides, using similar grasses as those recommended pond dikes (see Section 69, **Pond construction, 20/1**).

(d)   Cover the sides of the canal with **lengths of wood or bamboo** dri into the canal bottom next to each other.

(e)   Tied bundles of sticks or branches can also be used.

7.   More expensive ways include:

- **lining the canal** with bricks, cement blocks or concrete (Section 83, **Pond construction, 20/2**);
- lining the side slopes with elongated half-height **gabions** (Section 37, **Pond construction, 20/1**).

Note: any change in the quality of the canal walls will affect **the w discharge capacity** of the canal (see Section 82, **Pond construction, 2**

1. Fencing is often used on fish farms:

- to protect the growing stocks from theft; and
- to protect dikes, structures and plantations from animals.

2. Fences may also be used as **wind-breaks** (see Section 43), to diversify production (wood, fruit, leaves), to provide privacy and to improve the appearance of your fish farm.

3. There are **different types of fences**. You should select the right type according to its main purpose and the amount of money you wish to invest. The chart below should assist you.

Protection given by various fences

Type of fence	Relative cost[1]	Protection from	
		Animals	Theft
Live fence	2	Yes	No
Piled fence	1	Yes	Possible
Woven fence	2	Yes	No
Post-and-rail fence	3	Yes	No
Wire fence	4	Yes	No
Wire-netting fence	5	Yes	Yes
Stone wall	2	Yes	No

[1] From cheapest (1) to most expensive (5)

Live fences

4. **A live fence** is made of shrubs and trees planted closely together as a hedge and regularly trimmed to produce a barrier to keep animals out.

5. If planting material is readily available, these fences are quite **cheap to install**, needing only to be planted. They also have the advantage of providing **additional benefits** such as the production of wood, fruit and animal fodder as well as acting as low wind-breaks and being an attractive feature in the landscape.

6. However, before considering the use of live fences, you should be aware that:

- they usually grow quickly, **requiring labour** for regular cutting, sometimes twice a year;
- to be really stock-proof, they may require **the addition of some other type of fencing** at particular points; and
- hedges require more **space** than some other types of fence.

7. Several different **kinds of shrubs and trees** can be used to make live fences. If the availability of **animal fodder** for browsing is one of the major benefits you are interested in, you should use a shrub or tree **legume** which is fast-growing and resistant to grazing and trimming, such as *Cassia siamea* (Siamese senna) or *Gliricidia sepium* (Nicaraguan shad). For protection both from animals and theft, **a thorny live fence** can be effective. You could also use **agave** planted in alternate rows for the production of sisal (*A. sisalana*) or maguey (*A. cantala*). **Cactus** plantations are another possibility.

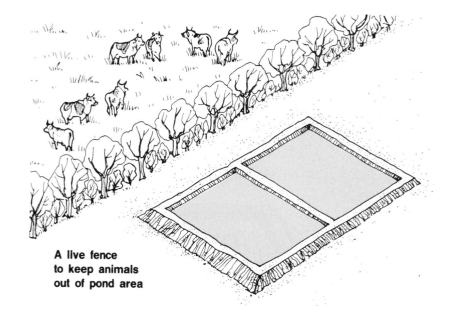

A live fence to keep animals out of pond area

Piled fences

8.   **A piled fence** is made of a row of cut vegetal material piled up high enough to keep out animals and sometimes also people. The best way is to use **thorny branches**, but you could also use any waste from land clearing or tree felling. Tie the material intermittently to vertical posts to give it more strength.

9.   Usually, this type of fence can last for some years, although it can also be very susceptible to fires, rot and termites.

**A piled fence
of thorny branches
or vegetal material
with intermittent
vertical posts**

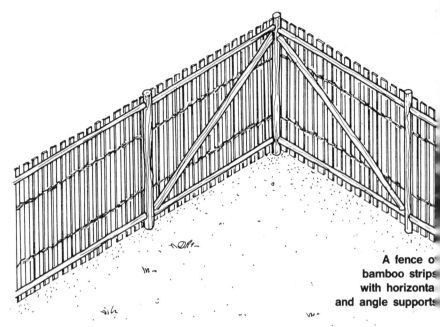

**A fence of
bamboo strips
with horizontal
and angle supports**

Woven fences

10. A simple light fence can be made using woven matting, for example bamboo, rush or leaf matting, fixed on a framework of wood or bamboo poles, with horizontal support bars at the top and bottom. These fences should be securely braced with angled support to keep them upright.

1. **A post-and-rail fence** consists of wooden posts strongly driven into the ground at regular intervals and joined with horizontal wooden rails. It particularly useful in areas where wood or bamboo is available.

2. This type of fence is easily made as follows:

) Prepare strong **wooden posts** with a diameter of at least 12.5 cm and 1.8 to 2.6 m long. If necessary, treat the wood to increase its resistance to termites and rot (see Section 31, **Pond construction, 20/1**).

(b) Prepare the **wooden rails**, usually using either lengths of split bamboo or wooden poles 10 cm in diameter.

(c) **Drive the posts vertically** into the ground to a depth of 0.5 to 0.8 m, at 2.5- to 4-m intervals. If necessary, add some angled supports to brace the posts.

(d) **Fix the rails to the posts**, in three to five horizontal rows. Make joints of the rails only at the posts, and not all on the same post.

(e) Install posts for gate openings. You may also want to install passes or stiles for crossing the fence (see page 127).

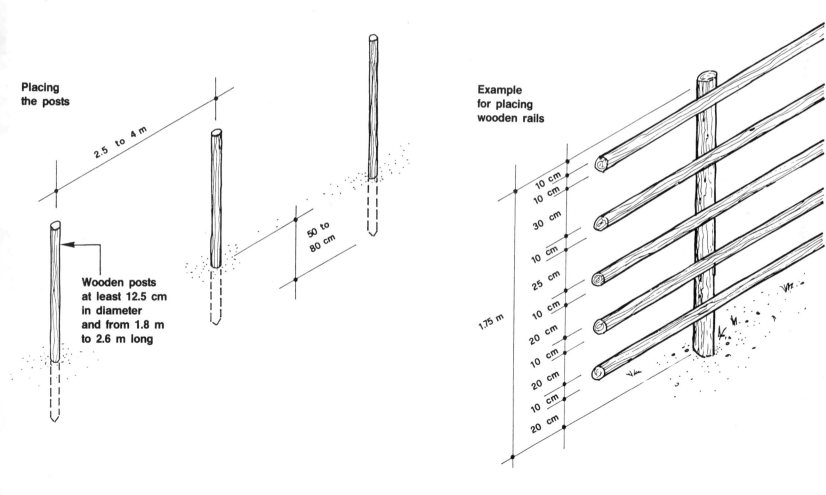

Placing the posts

2.5 to 4 m

50 to 80 cm

Wooden posts at least 12.5 cm in diameter and from 1.8 m to 2.6 m long

Example for placing wooden rails

1.75 m

10 cm
10 cm
30 cm
10 cm
25 cm
10 cm
20 cm
10 cm
20 cm
10 cm
20 cm

Plain wire gauges

Equivalence of common wire gauges (in mm)

Gauge	BWG	ISWG	USG	PG
5	5.588	5.384	5.314	1.0
6	5.154	4.876	4.935	1.1
7	4.572	4.470	4.554	1.2
8	4.191	4.064	4.176	1.3
9	3.759	3.657	3.797	1.4
10	3.403	3.251	3.571	1.5
11	3.047	2.946	3.175	1.6
12	2.768	2.641	2.778	1.8
13	2.412	2.240	2.381	2.0
14	2.108	2.032	1.984	2.2
15	1.828	1.828	1.786	2.4
16	1.650	1.625	1.587	2.7
17	1.472	1.421	1.428	3.0
18	1.244	1.218	1.270	3.4
19	1.066	1.016	1.111	3.9
20	0.888	0.914	0.952	4.4
21	0.812	0.812	0.873	4.9

Note: BWG = Birmingham Wire Gauge (UK)
ISWG = International Standard Wire Gauge
USG = United States Gauge
PG = Paris Gauge (France)

Data for Paris and Birmingham wire gauges

Wire gauge PG	BWG	Wire diam. (mm)	Wire section (mm^2)	Weight per 1 000 m (kg)	Length per kg (m)
5	20	1.0	0.785	6.12	163.40
6	19	1.1	0.950	7.41	134.95
7	18	1.2	1.130	8.81	113.50
8	18	1.3	1.327	10.35	96.62
9	17	1.4	1.539	12.00	83.33
10	17	1.5	1.767	13.78	72.57
11	16	1.6	2.011	15.68	63.77
12	15	1.8	2.545	19.84	50.40
13	15	2.0	3.142	24.48	40.85
14	14	2.2	3.801	29.64	33.74
15	13	2.4	4.524	35.28	28.34
16	12	2.7	5.725	44.63	22.40
17	11	3.0	7.068	55.13	18.14
18	10	3.4	9.079	70.82	14.12
19	9	3.9	12.045	93.17	10.73
20	8	4.4	15.205	118.59	8.43
20	7	4.6	16.619	129.62	7.71
21	7	4.9	18.857	147.08	6.80

3. A wire fence is made of **several wire lines** tightly stretched between series of **vertical posts**. Wire fencing is expensive and should be limited to areas where it is necessary and where cheaper fences cannot be established. There are two **kinds of wire**, usually **galvanized** for protection against corrosion:

- **plain wire**, which is cheaper but more difficult to place properly;
- **barbed wire**, which is more difficult to handle, rusts faster and can be a cause of serious injury to animals. It is most useful as a top wire.

Plain wire

Barbed wire

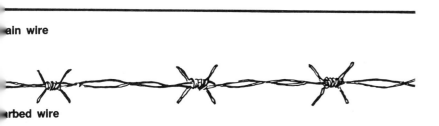

Plain wire is defined by its gauge (see the chart on page 120), while barbed wire quality is defined by **three numbers** defining the gauge, number of barbs (2 or 4) and distance between barbs (8 to 11 cm); for example, "16 × 4 × 11" describes a 16-gauge wire with four barbs every 11 cm.

Wooden posts are generally cheaper than concrete posts but less resistant to rot, fire and termites. Select a naturally durable hardwood or at less durable wood with a preservative (see Section 31, **Pond construction, 20/1**).

There are **two types of post** in a wire fence:

- **strainer posts**, at gates, extremities of straight lines, intermediate points on long lines and wherever the ground slope changes; and
- **intermediate posts**, at regular intervals between strainer posts.

17. The stability and efficiency of the wire fence will depend on **the strength and stability of the strainer posts**. You should therefore select and fix them with particular care.

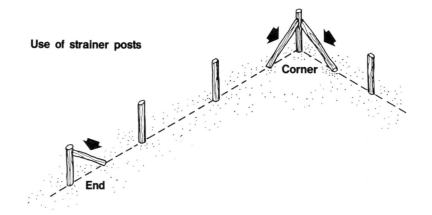

Use of strainer posts

Corner

End

18. Instead of standard posts, you can also use **live posts** consisting either of trees growing on the fence line or of specially planted trees. They are cheap and long-lasting, **providing additional benefits** such as wood, fruit, animal fodder and shade. Excellent live posts are made of *Gliricidia sepium*, a tree legume widely used at low altitudes in Latin America, where it is trimmed at 1 to 1.5 m in height to be directly grazed by livestock.

Note: trees to be used as live posts should be planted well in advance to give them enough time to take root properly.

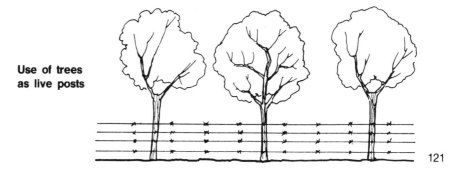

Use of trees as live posts

19. To make a simple wire fence, proceed as follows.

(a) **Lay out the fence line** as straight and unbroken as possible.

(b) **Clear the ground area**.

(c) **Install the strainer posts**, about 2 m long, at gates, passes, corners, top and bottom of slopes and at any intermediate point if the straight distance is over 150 to 200 m. They should be at least 15 cm in diameter and set into the ground 80 cm deep. Brace them well according to their position:

- brace **corner posts** with a stay in the direction of both fencing lines;
- brace **end or gate posts** in the direction of the fencing line;
- brace **intermediate strainer posts** at various points along long runs of fencing.

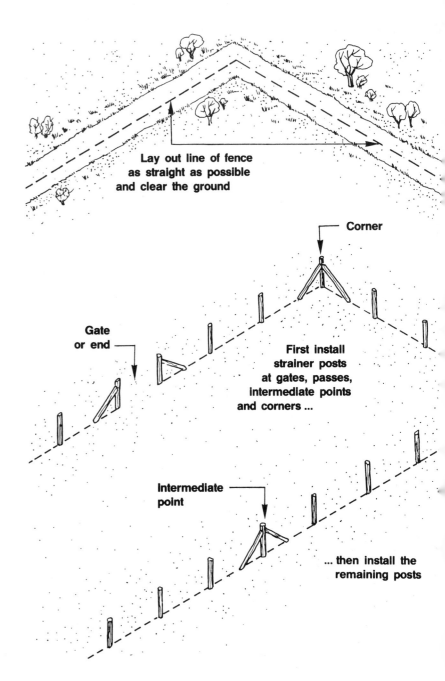

Lay out line of fence
as straight as possible
and clear the ground

Corner

Gate
or end

First install
strainer posts
at gates, passes,
intermediate points
and corners ...

Intermediate
point

... then install the
remaining posts

20. Different **types of stays** can be used according to soil conditions:

- single or double diagonal stays to stabilize the post by pushing against a surface or buried support at a 30° to 40° angle;
- a traditional strainer assembly with a diagonal stay between two posts;
- a strainer assembly with a horizontal stay and a double-twisted wire stretched between two posts.

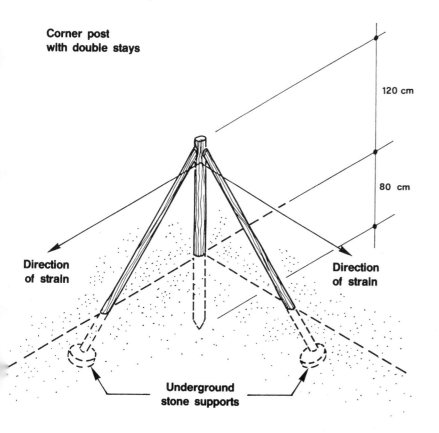

Corner post with double stays

120 cm

80 cm

Direction of strain

Direction of strain

Underground stone supports

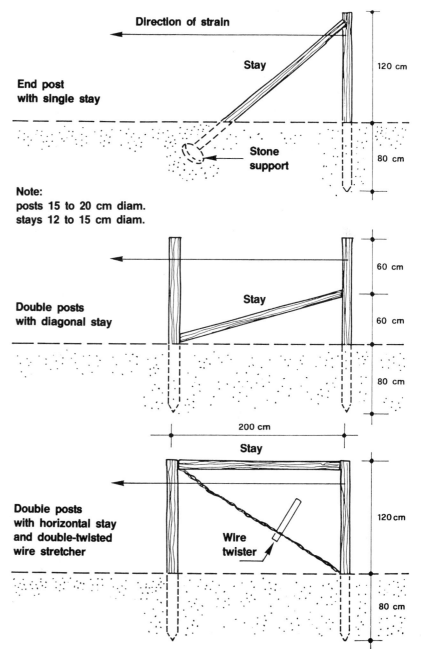

Direction of strain

Stay

120 cm

End post with single stay

Stone support

80 cm

Note:
posts 15 to 20 cm diam.
stays 12 to 15 cm diam.

Double posts with diagonal stay

Stay

60 cm

60 cm

80 cm

200 cm

Stay

Double posts with horizontal stay and double-twisted wire stretcher

Wire twister

120 cm

80 cm

123

21. You can either set these posts in dug or bored holes with firmly tamped soil around them, **or you can drive pointed posts into the soil**, giving them more stability. For this, it may be useful to build **a simple hand driver** as follows:

(a)   Get a steel pipe 20 cm in diameter and 90 cm long.
(b)   Close its top end by welding on a steel plate 25 cm × 25 cm.
(c)   At the other end, weld two handles.

22. The posts should be held vertically by a helper, while someone standing on a raised platform such as the back of a truck or a mobile wooden bench strikes the post head from above with the hand driver.

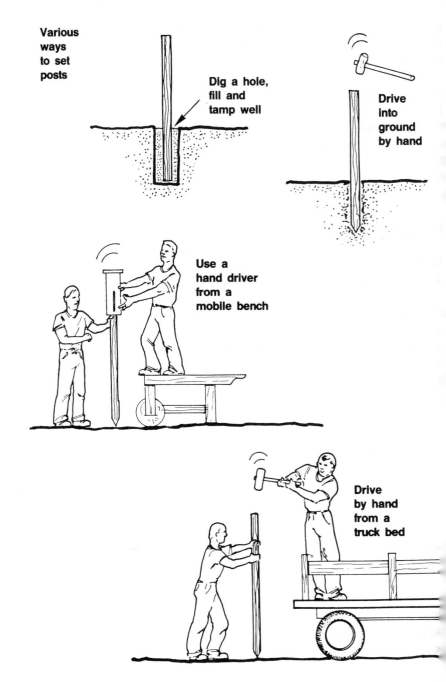

Various ways to set posts

Dig a hole, fill and tamp well

Drive into ground by hand

Use a hand driver from a mobile bench

Drive by hand from a truck bed

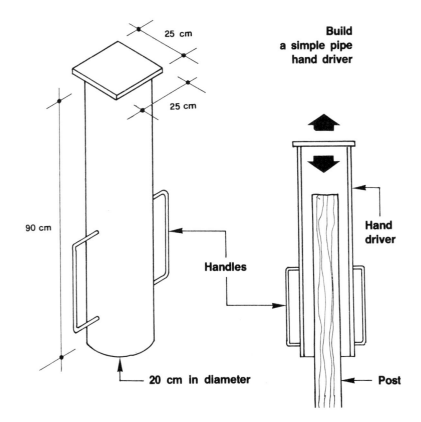

25 cm

25 cm

90 cm

20 cm in diameter

Build a simple pipe hand driver

Hand driver

Handles

Post

23. **Install the intermediate posts**, about 1.6 to 1.8 m long, exactly in line, at 3- to 5-m intervals. Their diameter should be 7.5 to 12.5 cm, and they should be set 40 to 60 cm deep.

24. **Install the bottom wire** 15 to 40 cm above the ground, securing it first to **a strainer post** and leading it straight to the next **strainer post**. It is then **tightly stretched** using a wooden lever and secured to the post. Make sure that **the wire passes on the inside of all the intermediate posts**. Make one turn around any **corner post** and secure the wire by wrapping it around and stapling it to the post. **Loosely fasten** the wire to the intermediate posts by stapling.

25. **Install the other three to five wires** from bottom to top, at 20- to 30-cm intervals, as described above. The top wire should be 10 cm below the head of the posts. Then **collect** all bits of wire, nails and staples left on the ground to keep animals from swallowing them later.

26. When all of the wire has been installed and stretched, it is time to **install gates**. Points for pond workers to cross fences, such as **passes**, should also be installed during construction. However, such crossing points as **stiles** can be placed after the fences are completed. Three kinds of gates, two kinds of passes and a stile are illustrated on pages 126 and 127.

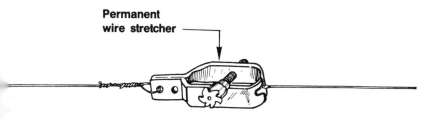

Permanent wire stretcher

Note: it is easier to stretch the wire if **permanent wire stretchers** are installed on it every 20 to 25 m; final stretching is done after the wire has been secured at each end to stretcher posts.

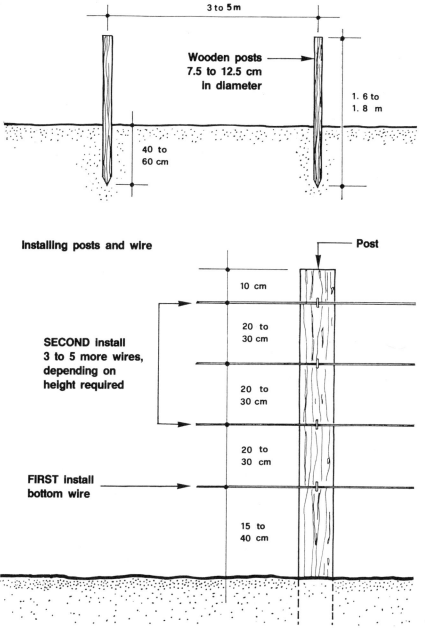

3 to 5 m

Wooden posts 7.5 to 12.5 cm in diameter

1. 6 to 1. 8 m

40 to 60 cm

Installing posts and wire

Post

10 cm

20 to 30 cm

SECOND install 3 to 5 more wires, depending on height required

20 to 30 cm

20 to 30 cm

FIRST install bottom wire

20 to 30 cm

15 to 40 cm

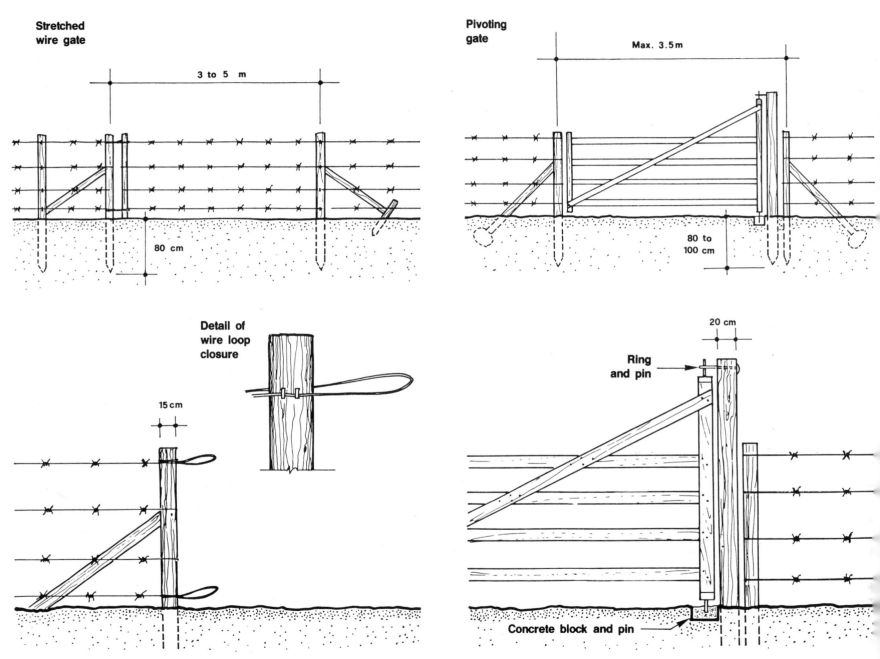

Stretched wire gate

3 to 5 m

80 cm

Detail of wire loop closure

15 cm

Pivoting gate

Max. 3.5 m

80 to 100 cm

20 cm

Ring and pin

Concrete block and pin

126

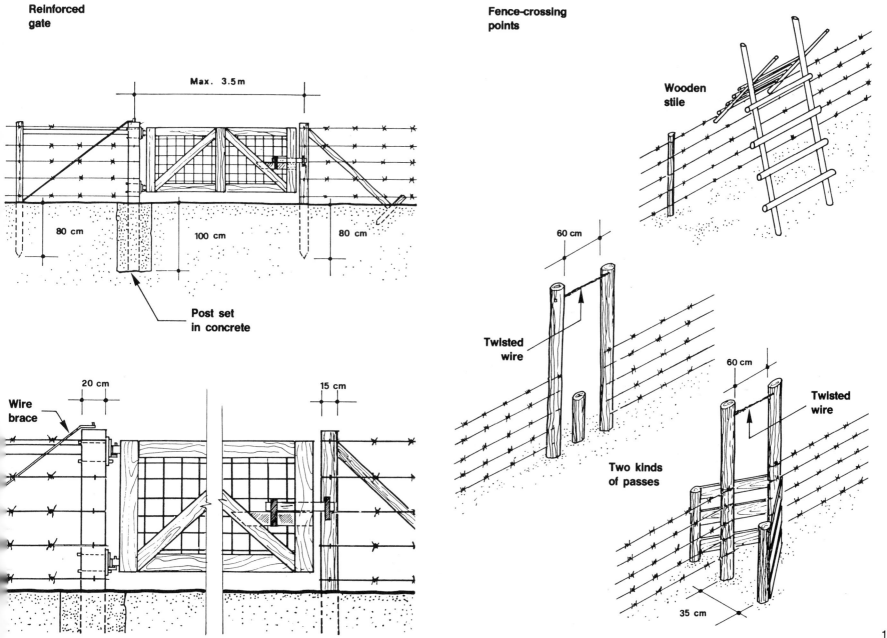

Reinforced gate

Max. 3.5 m

80 cm

100 cm

80 cm

Post set in concrete

Wire brace

20 cm

15 cm

Fence-crossing points

Wooden stile

60 cm

Twisted wire

60 cm

Twisted wire

Two kinds of passes

35 cm

127

27. Wire-netting fences are primarily used on fish farms to stop intruders and to protect the fish stocks from theft. To be effective, these fences should be high, dense, sturdy and topped with barbed wire. They can be quite expensive and **their use should be limited** to enclosing particularly valuable stocks such as broodstock and heavily stocked fish kept in storage ponds before being sold.

28. **Strong diamond-mesh netting**, about 2 m high and fixed to posts set 3 m apart, is commonly used for this type of security fence. Stays may be used for extra strength and barbed wire added at the top for increased efficiency. If possible, the barbed wire should be angled out to make access more difficult.

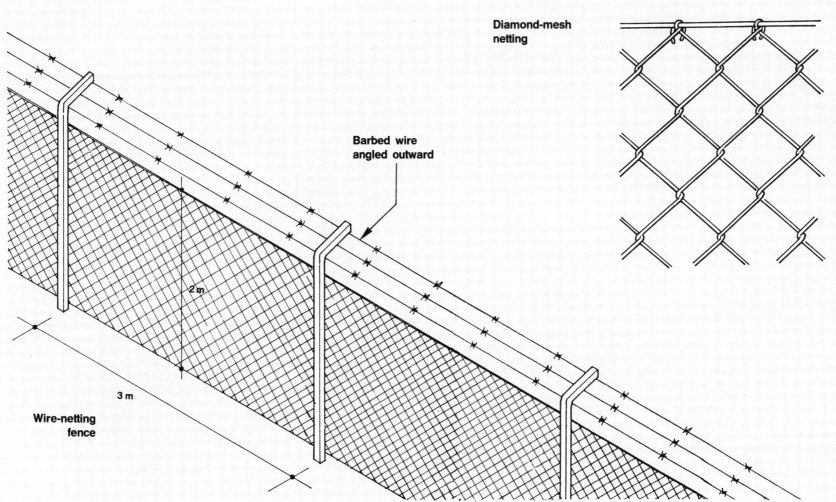

Diamond-mesh netting

Barbed wire angled outward

2 m

3 m

Wire-netting fence

9. **Whenever stones are plentiful**, a cheaper fence may be built by piling stones "dry" into **a wall** 0.7 to 1.2 m wide at the base. Although construction is labour intensive, maintenance costs are quite low. For extra security, short wooden posts can be built into the wall to carry some additional rows of barbed wire.

Stone wall

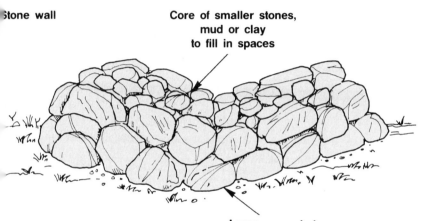

Core of smaller stones,
mud or clay
to fill in spaces

Larger, rounded
uncut stones
on the outside

Providing controlled access for livestock

30. In some integrated farms, livestock should be given access to at least **one pond for drinking**. It is then best to limit access to **a small portion of the pond**, which should be well protected against erosion with gravel, concrete or asphalt. You can use fences, as described earlier, to restrict access.

Controlled access

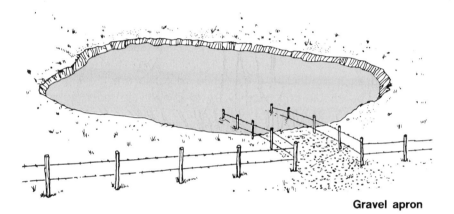

Gravel apron

46 Animal pest control in drained ponds

1. Farmed fish have many enemies and competitors, such as wild fish, frogs, insects and birds, from which they should be protected.

2. Protection is particularly important while the fish are still very small, for example still in nursery ponds. You can protect your fish in various ways.

(a) You can control the pests **after each complete harvest and before restocking the ponds**, the choice of method depending on whether:

*   you **can drain** your pond completely (see Section 46);
*   you **cannot drain** your pond completely (see Section 47).

(b) **During each production cycle** while your fish are growing in the pond, you should continuously try to control the most dangerous pests (see Section 48).

Controlling pests when the pond is drained

3. Pest control in drained ponds, also called **pond disinfection**, has several objectives, namely:

*   to kill **aquatic animal predators**, such as carnivorous fish, juvenile frogs and insects left in the water puddles and in the mud, which would survive and feed on the young fish to be stocked;
*   to eliminate all **non-harvested fish**, which later would compete with your new stock for space and food, especially if they reproduce without control;
*   to destroy **fish parasites and their intermediate hosts**, such as snails, and thus help **control diseases** (see Section 152, **Management, 21/2**).

4. Certain disinfection treatments have additional benefits such as improving water and bottom soil quality (see Section 50) or increasing the pond fertility (see Section 60).

Pest control in drained ponds

Animal pests and predators

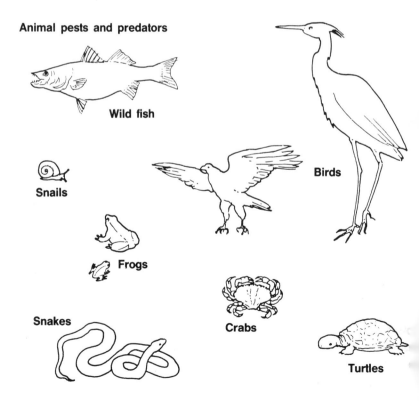

Wild fish

Snails

Frogs

Birds

Snakes

Crabs

Turtles

5. Earthen fish ponds are most easily disinfected **after their water has been drained** as thoroughly as possible, either by gravity for drainable ponds or by pumping for undrainable ponds. If necessary, **complete the draining** of remaining pools by deepening additional small trenches toward the main draining trench.

6. There are several ways to disinfect a drained pond. These are **usually combined** to give most reliable results.

(a) **Keep the pond dry** (preferably in warm, sunny weather). The ultraviolet rays of the sun have a powerful sterilizing effect. Depending on air temperature, keep the pond fully dry from 24 hours (at the minimum) to one month.

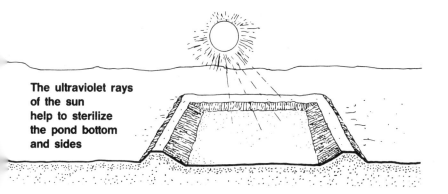

The ultraviolet rays of the sun help to sterilize the pond bottom and sides

Duration of dry period according to air temperature

Air temperature (°C)	Minimum duration of dry period
30	24 hours
20	4 days
Less than 20	30 days

Note: you learned earlier how to keep a pond dry to improve water quality (see Section 25). Remember that **certain types of pond should not be kept dry for too long**.

(b) **Throw or pour a toxic chemical** (see below) **in water puddles and pools** remaining on the pond bottom. Let the chemical react for a few days. Common procedures are summarized in **Table 9**.

Throw or pour a toxic chemical in puddles and pools

(c) Spread **a toxic chemical uniformly on the wet pond bottom and sides**. Let the chemical react for two to three days. Slowly let just enough water flow in to cover the whole pond bottom with a shallow layer. Wait for 15 days. Then drain slowly, taking care not to cause fish to be killed downstream. Before final refilling, it may be necessary to rinse out your pond once.

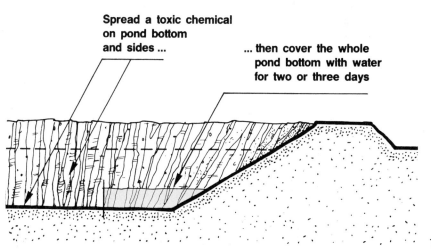

Spread a toxic chemical on pond bottom and sides ...

... then cover the whole pond bottom with water for two or three days

TABLE 9

Common disinfectants for totally drained ponds

	Doses		Remarks
	(g/m^2)	(kg/ha)	
QUICKLIME			Very caustic
Pond slightly muddy	100-200	1 000-2 000	Lumps or powder
Pond very muddy	200-300	2 000-3 000	Lime milk = 1 part + 4 parts water
HYDRATED LIME			Caustic, fine powder
Pond slightly muddy	160-300	1 600-3 000	To be freshly made
Pond very muddy	300-460	3 000-4 600	
CALCIUM CYANAMIDE			Highly toxic powder
Pond slightly muddy	100-150	1 000-1 500	Usually limited to treatment of small ponds
Pond very muddy	150-250	1 500-2 500	Rinse pond after use. Contains 60-70% CaO and 18-22% nitrogen
AGRO-INDUSTRIAL BY-PRODUCTS	To be used on flooded pond bottom or in puddles/pools, see **Table 10**		

Note: for a thorough disinfection, it is best to treat the drained pond twice at eight- to 15-day intervals. Refill the pond ten to 15 days after the second treatment.

Three chemicals are commonly used for disinfecting drained earthen ponds (see **Table 9**):

- **quicklime**, CaO (calcium oxide), sold in lump or powder form; it is **extremely caustic** and should be handled with great caution (see below). Because of its lasting toxicity, **lumps of quicklime should be well crushed** before use, but uniform spreading remains difficult. **It is better to use lime milk, prepared when needed** by mixing one part of quicklime lumps with four parts of water;
- **hydrated or slaked lime**, $Ca(OH)_2$ (calcium hydroxide), a fine powder also to be used with caution because it contains at least 65 percent CaO (buy and use only if freshly made);
- **calcium cyanamide**, $CaCN_2$, a powder containing 60 to 70 percent CaO and 18 to 22 percent nitrogen; it is highly toxic for several months after its application. Particular care should be taken when draining the solution from the pond to avoid killing fish downstream. Rinse out the pond well before final refilling.

Some agricultural by-products can also be used to disinfect drained ponds cheaply whenever they are locally available, for example **rice bran** (400 to 1 000 kg/ha), **crude sugar molasses** (400 to 500 kg/ha) and **tobacco dust** or tobacco **shavings** (300 kg/ha). Spread the required amount of by-product over the pond bottom. Flood with 5 to 10 cm of water for ten to 15 days. It is best not to drain the pond but to fill it up, so as not to lose the fertilizing effect of the organic disinfectant. Before stocking fish, check carefully on the dissolved oxygen content (see Section 25).

Note: before applying **tobacco dust or tobacco shavings**, it is best to **soak the sacks in water overnight**. This step will prevent the dust being blown away by wind during spreading on the pond bottom.

Taking precautions when handling dangerous chemicals

9. When using caustic and toxic chemicals, you should be very careful on your own behalf and on that of other people. Take the following precautions.

(a) Whenever possible, choose **a day without wind** to apply the chemical. If you have to do it on a windy day, progress in the direction of the wind, avoiding chemicals being blown over you or over others.
(b) **Fully protect skin and eyes** from contact with chemicals by using impermeable clothing, boots, goggles or glasses and a hat.
(c) **Avoid inhaling chemicals**. Protect mouth and nose with a piece of cloth.
(d) Thoroughly **wash your hands** before touching any food.
(e) Thoroughly **clean all equipment and clothing** when you have finished treating the pond.
(f) **Store all chemicals** so as not to present any danger to animal or human life, children in particular.

When using caustic or toxic chemicals, try to choose a day without wind ...

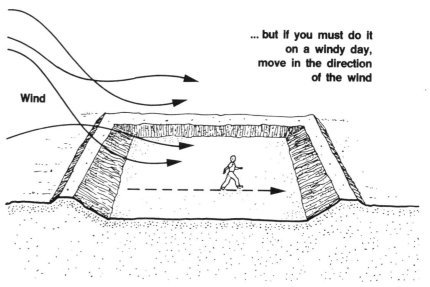

... but if you must do it on a windy day, move in the direction of the wind

Wind

47 Animal pest control in undrained ponds

1. **If the fish pond cannot be completely drained**, for example if it is undrainable or if there is no water available to refill it for the next production cycle, pests such as wild fish, frog eggs and tadpoles and snails should be controlled after harvesting the fish **by treating the water left in the pond**. **In nursery ponds**, voracious insects and even some zooplankters* should also be destroyed before restocking.

Controlling pests by treating the water

2. After harvesting your fish, proceed as follows to control the pests.

(a) **Lower the water level** as far as possible, keeping in mind that you will thereafter need water to refill the pond.

(b) **Estimate the volume** (in m³) **of water** present in the pond by multiplying its surface area (in m²) by its average depth (in m) (see Section 20, **Water, 4**).

(c) **Treat the water** with one of the products suggested in **Table 10**, ensuring that it completely mixes with the pond water. It should kill all fish, frog eggs and tadpoles, snails and most insects.

(d) After ten to 12 hours **collect the dead fish** with nets.

(e) **Wait ten to 15 days** for the organic poison to break down and disappear.

(f) **Stock a few fish**, preferably keeping them in a submerged net or a small cage to be able to watch them carefully.

(g) If these test fish survive well and do not show any abnormal reactions, your pond is **ready for stocking**.

Note: a similar procedure is also used in addition to netting as a means of harvesting fish from undrainable ponds (see Section 111, **Management, 21/2**).

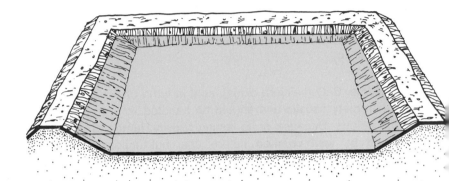

Pest control in undrained ponds

Applying powdered oilseed cake in an undrained pond

TABLE 10

Underwater pest control in ponds

Product	Doses
Quicklime	
little water in pond[1]	500-900 g/m^3
full pond[1]	200-250 g/m^3
Rotenone powder: usually 5% rotenone from	
Derris roots	20 g/m^3 [2]
Saponin, pure glycoside from plant	2-5 g/m^3
Derris root: from *Derris* spp., tuba; roots contain rotenone	20-40 g/m^3
Tephrosia leaves: leguminous tree; leaves contain rotenone	500 g/100 m^2
Barringtonia accutangula: powdered seeds	20 g/m^3
Croton tiglium: powdered oilseed cake	5 g/m^3
Milletia pachycarpa: powdered roots	5 g/m^3
Walsura piscidia: powdered bark	10 g/m^3
Bassia latifolia: oilcake; mahua (India)[4]	250 g/m^3
Camellia spp.: teaseed cake; 10-13% saponin[3, 4]	50-70 g/m^3

[1] Additional benefit: liming of water, see Section 52
[2] Equivalent to 1 g/m^3 pure rotenone
[3] Makes pond water acid; best if followed by quicklime treatment (15 g/m^3) to improve water quality, see Section 50
[4] Additional benefit: organic fertilization

3. Some of the products listed in **Table 10** require **processing before being used** in the pond:

(a) **Derris root**: select fresh, small roots. Cut into small pieces. Soak in water overnight. Pound, crush and squeeze to extract rotenone; dilute and mix with water in pond.

(b) **Teaseed cake**: dry and finely grind the seeds. Let them soak in lukewarm water for 24 hours. Dilute and mix with water in pond.

(c) **Rotenone or saponin**: mix the total required amount in water. Treat the pond while keeping this solution well mixed.

Controlling insects and zooplankters in nursery ponds

4. Several **aquatic insects** such as water beetle adults and larvae, dragonfly nymphs and adult water bugs can attack and destroy many fish fry in your nursery ponds. Most aquatic insects are also competing for the food necessary for your small fish to grow. Some of the largest **zooplankton species** may also harm very young fish fry. You should protect them from these enemies by:

- keeping grassy **ditches clean** in the farm neighbourhood;
- **refilling your nursery pond with water** not more than two weeks before stocking, if it has been drained completely;
- **treating the water** of the nursery pond to eradicate insects and sometimes zooplankters, as described in items 5 and 6 following.

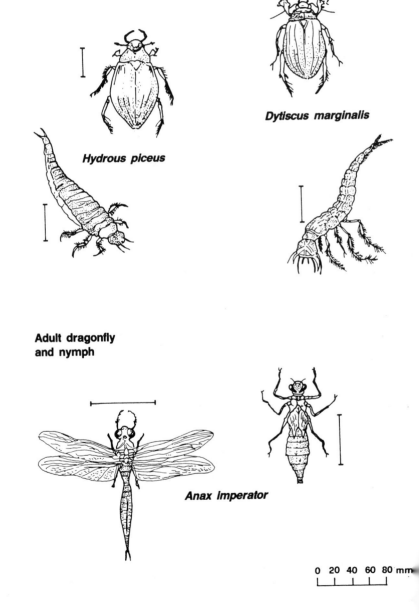

Adult water beetles and larvae (dysticid)

Hydrous piceus

Dytiscus marginalis

Adult dragonfly and nymph

Anax imperator

0 20 40 60 80 mm

Adult water insects

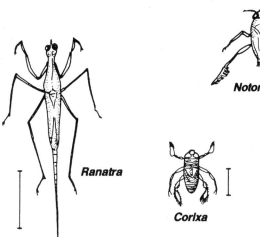

Notonecta

Ranatra

Corixa

Nepa

Small crustaceans (zooplankton)

Triops

Lepidurus

Limnadia

Branchipus

0 10 20 30 40 mm

5. **To eradicate aquatic insects which have to come to the water surface to breathe air**, such as the most harmful *Notonecta*, *Nepa*, *Ranatra* and dysticid species, proceed as follows.

(a) **Clear all the vegetation** from the pond. Cut the grass short on the wet dike slopes.

(b) Three to four days before stocking, slowly drag **a fine-mesh net** through the water to capture as many insects as possible and destroy them.

(c) Preferably on a calm, dry day but 12 to 24 hours before stocking, **spread a thin layer of an oily substance on the water surface** (see **Table 11**).

(d) Keep this layer undisturbed for **at least two hours**. If necessary, keep spreading more oily substance so as to **keep the layer unbroken** over the whole pond surface.

Note: the more wind there is, the more oily substance you will require and the less effective the treatment might be.

Dysticid larva and adult breathing at water surface

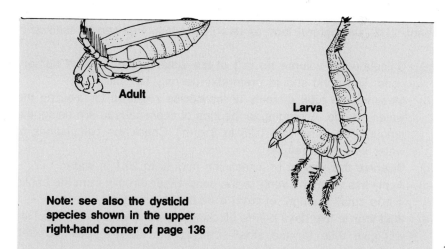

Adult

Larva

Note: see also the dysticid species shown in the upper right-hand corner of page 136

137

TABLE 11

Control of surface-breathing aquatic insects

Substance	Average dose per 100 m²
Diesel oil, high-speed quality	250-500 ml
Mixture 5:1 of diesel oil/used motor oil	250-500 ml
Mixture 4:1 of diesel oil/cottonseed oil	250-500 ml
Paraffin oil, kerosene or kerdane	200-400 ml
Mixture 3:1 of mustard oil/soap powder[1]	750 g
Mixture 3:1 of coconut oil/soap powder[1]	750 g

[1] Detergent or soap powder strengthens the floating layer

6.   **To eradicate all aquatic insects together with large zooplankters**, you will have to buy special agricultural chemicals. Most of these are organophosphates, **commercial insecticides** such as *Baytex*, *Dipterex*, *Dylox*, *Flibol*, *Fumadol*, *Masoten* and *Sumithion*. The last is particularly useful as it is toxic to insects, copepods and cladocerans but is not so toxic to small zooplankters such as rotifers (see Section 101, **Management, 21/2**), the natural food of very young fish. Proceed as follows.

(a)   Calculate **the volume (in m³) of the water** present in the nursery pond. The pond should preferably be only half full.

(b)   Measure **the total quantity of insecticide required** for treating the water volume. According to the kind of chemical you are using, the amount can vary from 0.25 to 3 g/m³. Check the calculations of water volume carefully.

(c)   **Dissolve** this weight of insecticide into 10 to 20 l of water.

(d)   **Apply this solution evenly to the pond** either directly from the banks if it is small enough, or from a boat if it is larger.

(e)   **Wait four to five days** before stocking the pond with young fry. This will give time for the small-sized zooplankters to develop well again.

48 Animal pest control in stocked ponds

1.   Once fish have been stocked in ponds, they should be protected as much as possible against **predators** such as carnivorous fish, frogs, turtles, snakes, birds and mammals. **Theft of fish** by people may also require preventive measures.

Controlling wild fish

2.   Once a pond has been treated for the extermination of all unwanted fish (see Sections 46 and 47), it should be protected from being reinfected through the water supply, through the use of one of **the filtering devices** described earlier (see Section 29).

3.   These filtering devices are particularly important for breeding and nursery ponds.

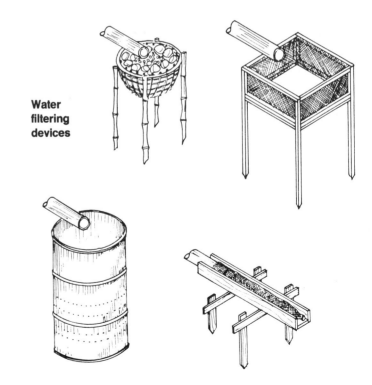

Water filtering devices

Trapping water turtles

Build a simple turtle trap as follows:

- Assemble **a square wooden frame** made of four boards. Add **four legs**. The frame should be high enough so that at least 30 cm of it will be above water level in the shallow part of the pond.
- Add a bottom made of wire mesh or chicken wire.
- Anchor the trap well in shallow water.
- Place a slanting board on the outside of this frame, leading from the pond bottom to its top edge. This will serve as a ramp for turtles.
- Fix another board at the end of the slanting board to tip inside the frame when weight is put on it. If you use spring hinges, the tip board will return to the upright position after a turtle has fallen into the trap.
- All around the upper edge of the frame, drive in a row of nails fairly close together and bend them slightly down.

Bait the trap at its centre with fish or meat. Any turtle climbing up one of the slanting boards and crawling on to the tip board should fall into the trap.

Trapping water snakes

A simple cylindrical trap, 25 cm in diameter and 70 cm long, can be made with fine metal netting. The two end openings are closed with funnels, one of which can be easily removed.

Install several traps **close to the area where the young fish are fed**. Half of the trap should be kept under the water surface. Bait the traps with dead fish or frogs and check them every day.

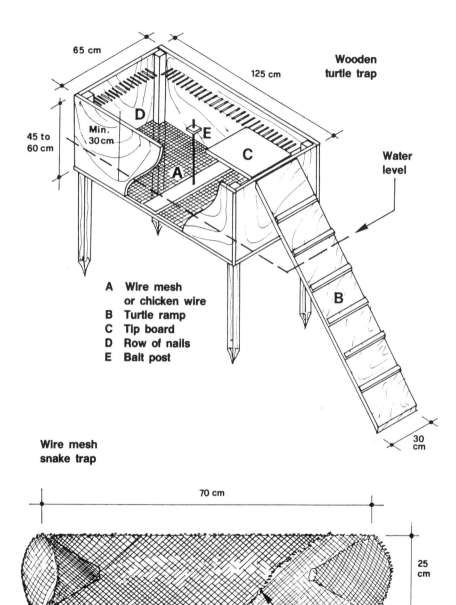

Wooden turtle trap

Water level

A Wire mesh or chicken wire
B Turtle ramp
C Tip board
D Row of nails
E Bait post

Wire mesh snake trap

Cutaway showing wire mesh funnel inside

139

Controlling frogs

8. Frogs are equally harmful in breeding and nursery ponds, where some species such as *Xenopus* actively feed on young fish; in addition, tadpoles compete for food.

9. Frog control is not easy and seldom completely successful. **To reduce the population** as much as possible, try the following regularly.

(a) **Catch tadpoles**, either with a scoop net or a lift net. Use the lift net over a feeding trough where the tadpoles accumulate, or bait the net itself with some food to attract tadpoles. A lift net can be attached to a movable post, which can be taken from place to place around a pond or from pond to pond.

(b) **Trap adult frogs**, especially *Xenopus*, in baited wire traps set along the pond banks. This is similar to the wire trap shown for snakes on page 139.

(c) **Fence those ponds** that need particular protection using a material over which frogs cannot climb, such as plastic or corrugated metal sheets. Such a protection should be at least 50 cm high.

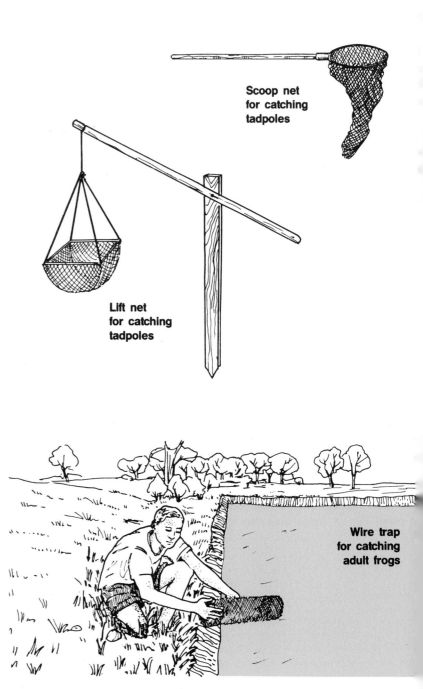

Scoop net
for catching
tadpoles

Lift net
for catching
tadpoles

Wire trap
for catching
adult frogs

Protecting against birds

Several kinds of bird are fond of fish, for example kingfishers, herons, fish eagles, pelicans and cormorants. In shallow ponds and wherever fish are concentrated, the damage and losses incurred can be considerable.

There are several ways to protect **particularly vulnerable areas** of the fish farm, such as broodstock, breeding, nursery and storage ponds. Those most commonly used are:

- **scaring devices** installed next to or inside the ponds, such as scarecrows, bamboo rattles, empty-can rattles, pieces of mirror on string, or flags;
- **thin wires** or strings stretched across the ponds and perpendicular to prevalent wind to keep the flying birds from landing on, sweeping over or diving into the ponds. On small ponds, stretch wires or strings 50 to 70 cm apart, on larger ponds, up to 8 to 15 m apart; attach streamers to the wires;
- **fencing the shallower parts** of the ponds or stretching thin wires about 50 cm above water level and 50 cm offshore to keep out wading birds;
- **covering the ponds** with netting material.

One of the best deterrents is **human activity** near the ponds. If you live on your fish farm you will have far fewer problems than if your ponds are completely isolated.

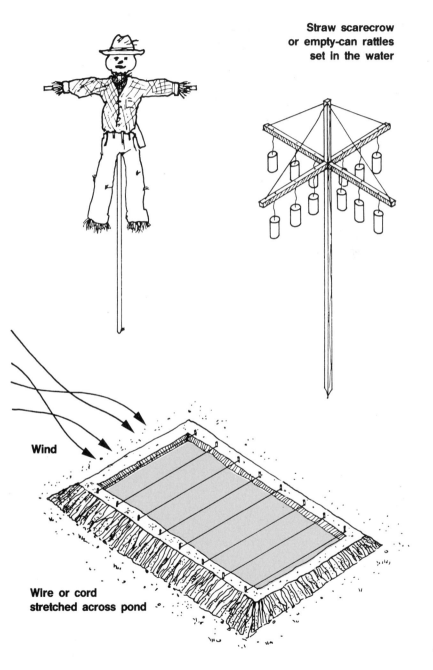

Straw scarecrow or empty-can rattles set in the water

Wind

Wire or cord stretched across pond

Controlling harmful mammals

13. Aquatic mammals can be harmful in two ways:

- some, like **otters**, can be active predators;
- others, like **muskrats**, can seriously damage pond dikes by digging long burrows above water level.

14. Apart from fencing, which might be expensive, the most effective control is obtained by:

- **trapping**, using specially designed traps; and
- **poisoning**, using poisoned baits.

15. Action should be taken as soon as it has been discovered that an animal has become active on the farm itself or in the neighbourhood.

Otter with fish

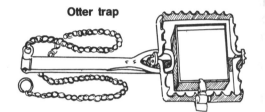

Otter trap

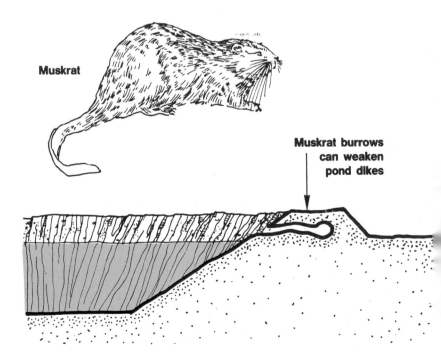

Muskrat

Muskrat burrows can weaken pond dikes

16. Theft of fish caused by people is unfortunately rather common on fish farms, particularly if the ponds are isolated and far away from housing.

17. As you learned earlier in Section 45, a good **security fence** is expensive. Its use should be limited to the protection of the most valuable fish stocks, such as broodstocks and food fish held ready for marketing.

18. To protect a fattening pond from **poaching with nets** such as seine or cast nets (see Sections 82 and 83, **Management, 21/2**), you can use one of the following simple ways.

a) Throw **bamboo, stripped of leaves, with plenty of lateral branches** into the pond, particularly along its banks.

b) Take a young tree and cut off all its branches, leaving 10 to 15 cm close to the tree trunk. Cut off the tip of the tree down to where the stem is about 15 cm thick. Cut this end into a point and **drive the tree upside-down** about 50 cm deep into the pond bottom. Repeat for several trees, placing some at 1.5 to 2 m from the banks and others in the rest of the pond area.

c) **Make some hooks** about 60 cm long from iron rods and set them into old metal containers filled with concrete. Sink these devices into your pond at various locations.

d) Install a series of **wooden stakes** in the pond bottom. Join them with **barbed wire** or simply secure a few layers of barbed wire around the head of each stake.

19. It is usually difficult **to steal fish by draining a pond**, unless it is unguarded for long periods of time. Avoid the risk, and avoid loss of water caused by someone tampering with **the outlet structure**. Make access to monks and pipes difficult, by removing valve handles and, if necessary, putting locks on monk and sluice boards.

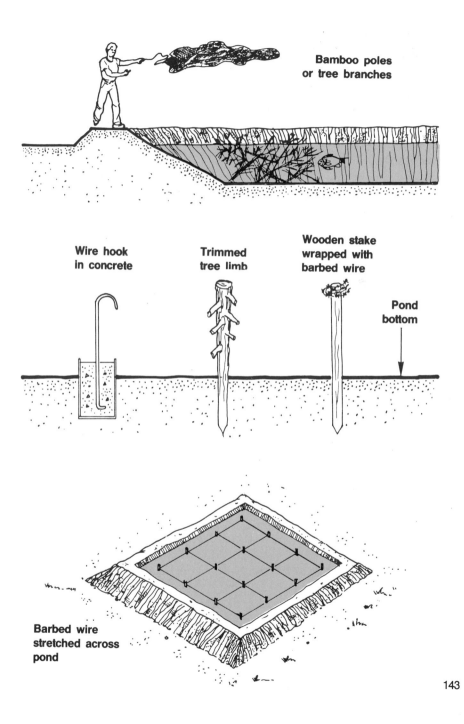

Bamboo poles or tree branches

Wire hook in concrete

Trimmed tree limb

Wooden stake wrapped with barbed wire

Pond bottom

Barbed wire stretched across pond

143

49 Aquatic vegetation control in fish ponds

Introduction

1. Aquatic vegetation may exist in various forms and is usually classified as floating, submersed or emersed plants. A small growth may not be harmful but, if it becomes excessive, it may result in adverse effects on fish pond management. **Vegetation control** then becomes necessary for several reasons:

(a) Vegetation absorbs too many **nutrients** from water and bottom mud.
(b) It reduces **sunlight penetration** into the water, reducing **photosynthesis** (see Section 20).
(c) It provides refuge for the **enemies and competitors** of fish.
(d) It makes **the pond bottom too rich** in cellulose fibres, slowing down the decomposition of mud.
(e) It may greatly hamper **the harvesting** of farmed fish.

2. Vegetation should be controlled regularly as part of **routine pond management**. It is cheaper to weed out vegetation before it becomes excessive. Natural food production and sanitary conditions will be improved as a result.

Remember: a fringe of emersed plants can be useful for protecting dikes exposed to the wind (see Section 43).

3. There are **three** ways to control aquatic vegetation:

- **biological methods**, using specific animals;
- **mechanical methods**, by hand or with the help of equipment;
- **chemical methods**, based on the application of specific chemicals.

4. As far as possible, you should choose biological or mechanical methods. **Chemical methods should be restricted** to situations where the other methods cannot be applied, for example because of high cost of labour or large farm size.

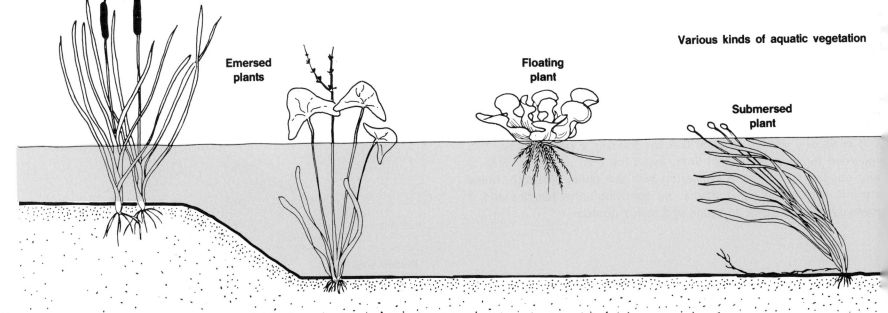

Various kinds of aquatic vegetation

Emersed plants

Floating plant

Submersed plant

Some herbivorous fish can be effectively used to control many kinds f plants, for example:

- Chinese grass carp (*Ctenopharyngodon idella*): use 100 to 200 g fish at the average stocking rate of 1.5 to 2 kg per 100 m² of water surface;
- *Puntius gonionotus*: usually stocked as fingerlings, 10 to 20 g at 2 to 5 kg/100 m²;
- *Tilapia rendalli* and *Tilapia zillii*: stocked as fingerlings of 10 to 20 g at 2 to 5 kg/100 m², or as larger fish (at least 50 g) at 1 to 2 kg/100 m².

Provided they are not too hard, most submersed, floating and even mersed plants are eaten by these fish. The bigger the fish, the larger d harder the plants they can usually control.

Some fish, such as the **common carp, can indirectly control vege- ion growth** by digging the bottom mud for food and increasing the ter turbidity. It has been observed that at least ten fish of 250 g or ore each are necessary per 100 m² of water surface. A similar result ay be obtained through **water fertilization**, which will increase plankton bidity (see Section 60).

Ducks can also help to control submersed vegetation in the shal- er parts of ponds while searching for food (see Section 73). They also d on some floating plants such as duckweed, *Lemna* sp.

Some kinds of fish and ducks can help to control plants

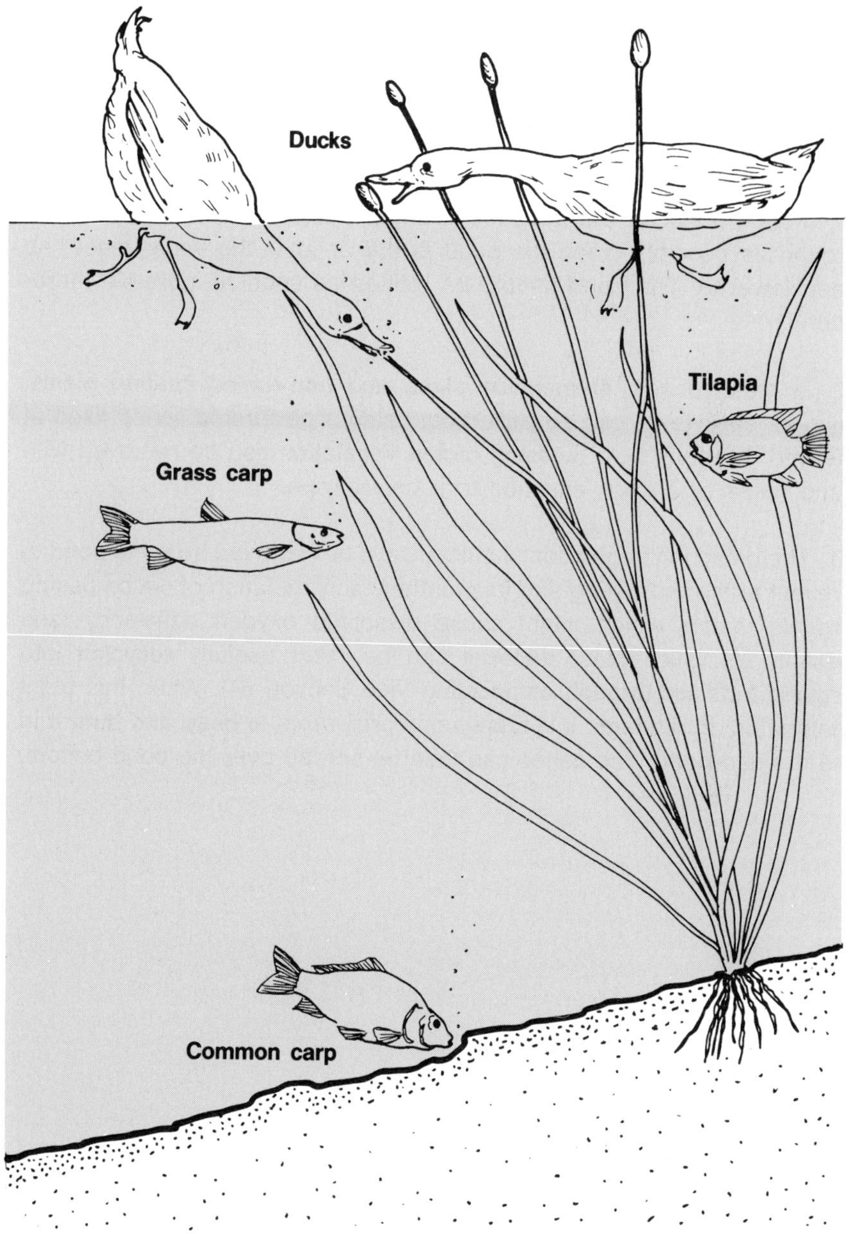

Ducks

Grass carp

Tilapia

Common carp

145

9. **Hand clearing** of aquatic vegetation is probably the most common method used. It can be most effective in smaller ponds, especially if carried out in drained ponds just before they are refilled.

10. It is best **to pull out rooted plants** entirely. With some strong emersed species such as reeds and papyrus, this is not always possible.

11. If the roots of the plant are too strong to pull, cut them as close to the ground as possible when the pond is dry or after the water level has been lowered. The pond should be refilled as soon as possible thereafter.

12. In small ponds, **filamentous algae** and **non-rooted floating plants**, such as duckweed, can be pulled out using a perforated scoop fixed at the end of a long pole. **Loosely rooted vegetation** can be removed with hand rakes or a hook attached to a heavy rope.

13. The detached vegetation should always be **removed from the pond** to prevent further sprouting and to avoid any accumulation of decomposing organic matter which might cause dissolved oxygen deficiency (see Section 25). Such plant material can be most usefully **recycled into organic fertilizer through composting** (see Section 64). When this plant material is dry enough, it is sometimes preferable **to heap and burn it** in the drained pond. The ashes can then be spread over the pond bottom.

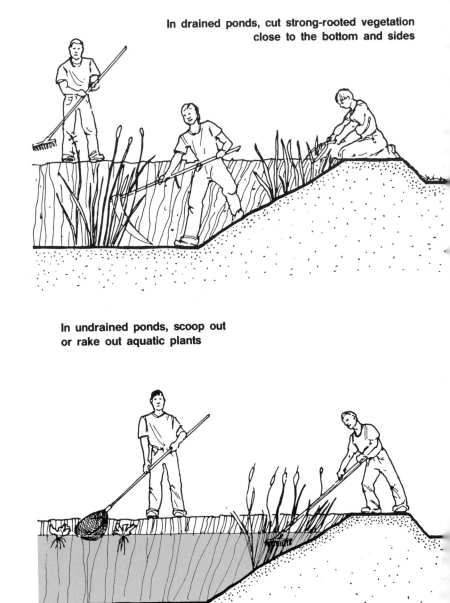

In drained ponds, cut strong-rooted vegetation close to the bottom and sides

In undrained ponds, scoop out or rake out aquatic plants

Controlling aquatic vegetation chemically

4. Several commercial chemicals are known to control algae (**algicides**) or higher plants (**herbicides**), in particular emersed and floating species. The control of submersed species is much more complicated, especially in full and stocked ponds.

5. **The control of algae and algal blooms** is best done with **copper sulphate** ($CuSO_4$), a very effective and cheap algicide which can be safely used in stocked ponds if the total alkalinity of the water (see Section 50) is sufficiently high. For best results, remember the following points:

a) Determine **the total alkalinity** of the water (see Section 50). If it is higher than 50 mg $CaCO_3$/l you can treat the pond in the presence of fish. If not, you have to harvest the fish first.

b) Determine which **type of algae** are to be controlled:

- for a microscopic **unicellular species** or **a larger colonial species**, use an average dose of 0.1 to 0.3 g of copper sulphate per m^3 of water;
- for a **filamentous species** such as *Spirogyra*, *Anabaena*, *Oscillatoria*, *Chara* or *Nitella*, you will require higher doses of copper sulphates. Proceed according to the chart on page 149.

Note: the bar scales adjacent to all examples of cellular, colonial and filamentous algae on this page and on page 148 are equal to **ten micrometres**, with the exception of the plant-like filamentous algae in the lower right-hand corner of page 148 where the bar scales are equal to **one millimetre**.

Remember: 1 micrometre (micron) = 1μm = 0.001 mm or one thousandth of a millimetre.

Examples of unicellular algae

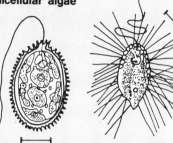
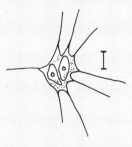
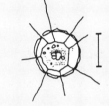
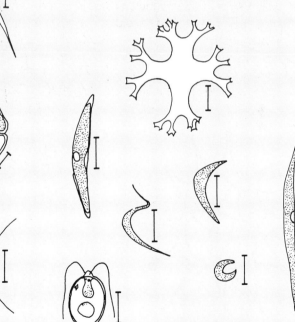

Note: bar scales = 10 μm

147

**Examples of
colonial algae**

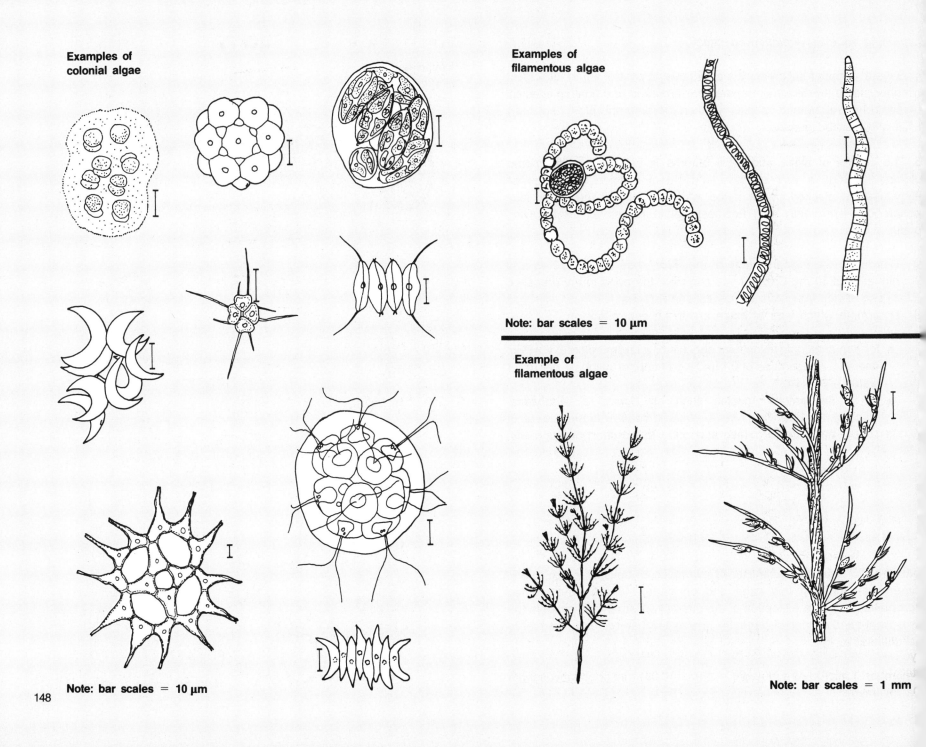

**Examples of
filamentous algae**

Note: bar scales = 10 μm

**Example of
filamentous algae**

Note: bar scales = 1 mm

Note: bar scales = 10 μm

148

Control of filamentous algae with copper sulphate

Water total alkalinity (mg CaCO$_3$/l)	Treatment
Lower than 50	Do not treat in the presence of fish During their absence, use an average dose of 0.5 g/m^3
50-100	Treat with caution in the presence of fish using an average dose of 0.5 g/m^3
100-200	Treat at the average dose of 0.5-1 g/m^3
200-400	Treat at the average dose of 1-2 g/m^3
Higher than 400	Do not use copper sulphate for control

Note: 1 g/m^3 = 1 mg/l

If fish are stocked, it is safer **to treat only one-third or one-half of the pond surface** at any one time, particularly if the total alkalinity of the water is low or if you are using concentrations higher than 1 g/m^3. It is most important **to determine carefully the volume of water** to be treated (see **Water, 4**).

Whenever possible, choose **a dry, calm day** for the treatment.

It may be necessary **to repeat the treatment** after a few weeks.

After the treatment and the killing of the algae, carefully **watch for any sign of dissolved oxygen deficiency** and take appropriate action (see Section 25).

Copper sulphate is a poison. **Handle and store it with care**.

If possible, test the dose of copper sulphate you plan to use on some your fish in two to three plastic or glass containers filled with pond ter, before treating the stocked pond. Observe the behaviour of the over 24 to 36 hours (see also Section 153, **Management, 21/2**).

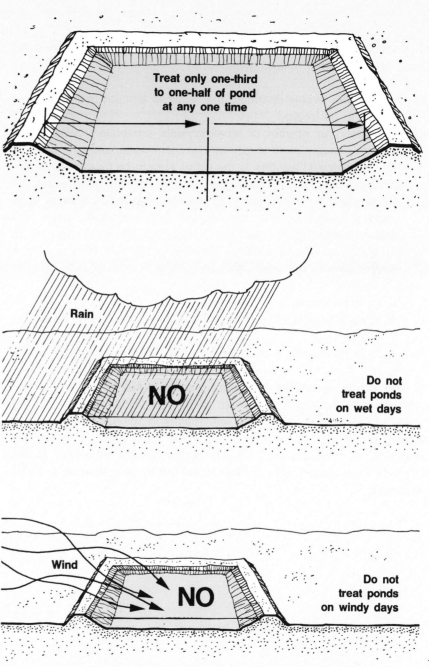

Treat only one-third to one-half of pond at any one time

Rain

NO

Do not treat ponds on wet days

Wind

NO

Do not treat ponds on windy days

17. Copper sulphate exists in two forms, powder or crystal. **To treat the pond**, use one of the following methods.

(a) **For powder**, evenly distribute the required amount over the water surface to be treated.

(b) Alternately, **for powder or small crystals**, completely dissolve the required amount in 10 to 20 l of lukewarm water and evenly apply this concentrated solution to the water area to be treated. Keep the solution well mixed. Ensure good mixing with the pond water.

(c) **For crystals**, put the required amount in one or more **small bags** made of mesh material, such as netting. Attach these bags to a **floating support**. Distribute the chemical in the pond area to be treated as evenly as possible by moving the float back and forth over this area. You may use:

- in small ponds, one inner-tube;
- in medium-sized ponds, two inner-tubes connected by a wooden support;
- in larger ponds, a boat with the bags attached over the sides.

18. **The use of herbicides to control aquatic higher plants** is a much more delicate matter. The success of the herbicide depends on the choice of the most suitable chemical mainly in relation to the kind of vegetation, the time of the year, the fish species present, the treatment method to be used and the care with which it is applied. In fish farms, herbicide treatment should only **be carried out by competent and skilled personnel**.

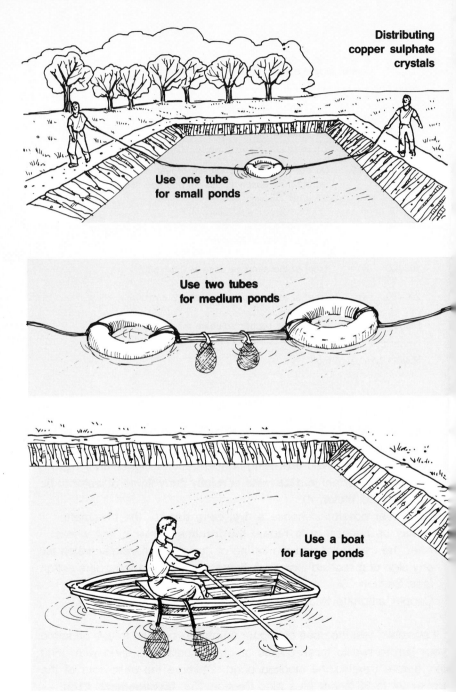

Distributing copper sulphate crystals

Use one tube for small ponds

Use two tubes for medium ponds

Use a boat for large ponds

5   POND CONDITIONING THROUGH LIMING

1. The aim of good pond management is to increase fish production through **an improved supply of natural food** such as phytoplankton and zooplankton. The supply is usually increased by fertilizing the pond water. You will learn more about this and the fertilizers to be used in Chapter 6. However, in this chapter you will first be told **how to prepare and**, if necessary, **treat your ponds so that these fertilizers will work as well as possible**. This preliminary step is called conditioning the ponds.

2. Earthen ponds are conditioned by **liming**, i.e. preparing the ponds and treating them with various types of **lime, chemical substances rich in calcium (Ca)** similar to those you have learned to use to control pests (see Section 46).

3. This process improves the structure of the pond soil, improves and stabilizes water quality and makes the fertilizing materials act more efficiently to increase food supply.

4. One of the most important effects, and one which you can measure and use to control liming, is the effect on the **total alkalinity** of the pond water.

Total alkalinity

5. **The total alkalinity (TA)** of water is a measure of its **total concentration in carbonates* and bicarbonates*** of substances, such as calcium (Ca) and magnesium (Mg), which are characteristically alkaline (see Section 22). In natural waters, **calcium bicarbonates** are usually predominant.

6. Total alkalinity has **great importance in fish farming**. It tells you how much the water pH can vary and how good the availability of carbon dioxide (CO_2) is, for example, for the production of microscopic algae (see Section 20). The toxicity of certain chemicals, such as copper sulphate, can vary according to the total alkalinity of the water (see Section 49).

7. Total alkalinity depends on the local characteristics of the soils and water, and on the way your farm is operated. It is affected directly by liming, which adds calcium material to the water and to pond soils.

8. Water with high alkalinity is also said to have **a good buffering capacity**. It is quite stable chemically, and its quality does not vary much through the day.

9. You may also have heard about **total hardness** of water, which is primarily a measure of the amount of **calcium and magnesium** present. In waters good for fish farming, total hardness does not greatly differ from total alkalinity. Thus, **soft water** (low hardness), with little calcium and/or magnesium, usually also has a low alkalinity, while **hard water** (high hardness) tends to have a high alkalinity.

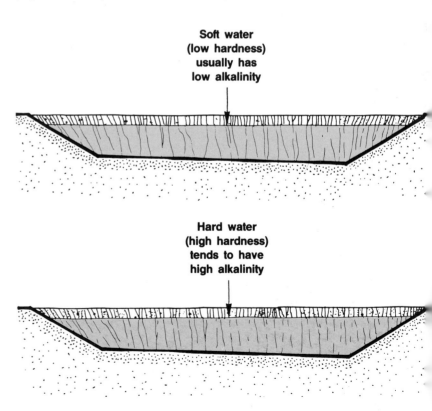

Soft water
(low hardness)
usually has
low alkalinity

Hard water
(high hardness)
tends to have
high alkalinity

Measuring total alkalinity

10. To measure the total alkalinity of water, you need two chemicals:

- a **0.1 normal** solution of hydrochloric acid (HCl);
- some methyl orange indicator solution.

11. Proceed as follows.

a) Obtain **100 ml** of the water to be tested.

b) Add **three drops** of the methyl orange solution to colour the water yellow.

c) Using a pipette graduated in millilitres and filled with **HCl solution**, slowly add this acid drop by drop while mixing the water sample well, **until its yellow colour turns orange-yellow**.

d) Confirm this by adding **one more drop**: the orange-yellow colour should now turn orange-pink.

e) Measure how many **millilitres of HCl solution** was used, for example A ml.

Expressing total alkalinity

12. Total alkalinity (TA) is usually expressed in one of two ways:

- as the **acid-binding capacity, in SBV units**, SBV referring to the original German name **S**aüre**b**indungs**v**ermögen (when the above measurement is completed, determine **TA = A** SBV units); or
- in **mg/l of equivalent calcium carbonate** ($CaCO_3$) (when the above measurement is completed, determine **TA = 50 A** mg/l $CaCO_3$).

Example

You measure A = 2.5 ml; the total alkalinity of the water is equal to 2.5 (SBV) or 2.5 × 50 = 125 mg/l $CaCO_3$.

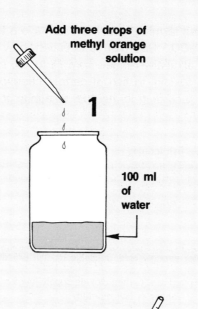

Add three drops of methyl orange solution

1

100 ml of water

Using a graduated pipette, add HCl solution drop by drop until the water turns orange-yellow

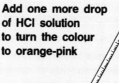

2

Add one more drop of HCl solution to turn the colour to orange-pink

3

A — Measure number of millimetres used

4

Using total alkalinity in fish farming

13. To be of some value for pond fish farming, water should have **a total alkalinity greater than 25 mg/l CaCO₃**. Best fish production may be obtained in waters where the total alkalinity ranges from 75 to 175 mg/l $CaCO_3$ (see the chart below). The toxicity to fish of certain chemicals, such as copper sulphate, can vary according to the total alkalinity of the water (see Section 49).

Alkalinity and fish farming

Total alkalinity of water		Potential for fish farming
in SBV units	in CaCO₃ mg/l	
<0.1	< 5	Very low: water strongly acid, unusable for fish breeding
0.1−0.5	5− 25	Low: water pH variable; carbon dioxide supply low for plant photosynthesis; danger of fish mortality
0.5−1.5	25− 75	Medium: water pH variable; carbon dioxide supply medium
1.5−3.5	75−175	High: water pH varies only between narrow limits; carbon dioxide supply optimal for plant photosynthesis, especially phytoplankton
>3.5	>175	Medium to low: water pH very stable; carbon dioxide supply decreases as alkalinity increases; fish health not endangered; calcareous deposits may form on surfaces

51 When to use lime to condition your ponds

Introduction

1. **Liming fish ponds is not always necessary**. In certain cases, it may not only be a waste of money but it can also be harmful to your fish. Before making any decision, you should carefully study your own ponds and their particular water and bottom soil characteristics. Check for the following:

(a) **If the pH of the pond bottom soil** is less than 6.5, liming of the bottom soil is justified.

(b) **If the pond bottom is very muddy** because it has not been regularly drained and dried, liming will improve soil conditions.

(c) If there is the danger of the spread of **a contagious disease** or **common pests should be controlled**, liming can help, especially in drained ponds (see Section 46).

(d) **If the amount of organic matter is too high**, either in the bottom soil or in the water, liming may be advisable.

(e) **If the total alkalinity of the water** is less than 25 mg/l CaCO₃, liming could be justified.

(f) **If the pH of the pond water at the end of the day is low** (see Section 22), the chart below shows the need for liming.

Liming desirability

Water pH	Liming of the pond water
<5.5	Obligatory
5.5−6.5	Necessary to increase pH and alkalinity
6.5−8.5	Eventually to increase alkalinity
>8.5	None/dangerous

2. **Liming will have little effect** and might be difficult to economically justify if:

- the bottom soil pH is above 7.5;
- the water exchange through the pond is too fast;
- the water pH at the end of the day is 7.5 or above;
- the water total alkalinity is above 50 mg/l CaCO₃.

Generally, **ponds should not be limed** if:

- **fertilizers** will not be used subsequently, unless the water is very acid;
- **natural food** is not important, the fish being fed a complete diet;
- the **water pH** reaches more than 8.5 by the end of the day.

beneficial effects of liming

If the above criteria justify treating your ponds with lime, **a series of beneficial effects** both on the bottom soil and on the water should result in increased fish production.

The **effects on bottom soil** are:

- **structure** will be improved (see **Soil, 6**);
- **decomposition of the organic matter** will be accelerated; and
- **pH** will increase.

All of these factors will result in a faster and greater release of minerals and nutrients from the bottom soil back into the pond water, together with a reduced demand for dissolved oxygen.

The effects on pond water are:

- **pH** will increase and become more stable;
- **total alkalinity** will increase, providing more carbon dioxide for photosynthesis;
- **calcium content** will increase, to be used by plants;
- certain **toxic substances** such as iron compounds will be neutralized and precipitated as pH increases; and
- excess **organic matter** will precipitate, decreasing the demand for dissolved oxygen in the pond water.

52 Chemicals for pond liming

1. **Three basic chemicals** are commonly used for liming fish ponds:

- calcium carbonate, $CaCO_3$;
- calcium hydroxide, $Ca(OH)_2$, or hydrated;
- calcium oxide, CaO, or quicklime.

2. Each of these produces a different **type of lime**; toxicity for fish, the effectiveness for pond liming and the cost will vary (see **Table 12**). Some of the other characteristics of various liming materials are given in **Table 9**.

3. **The efficiency of liming materials** increases as their individual **particle size decreases**. Before use, you should ensure that the lime is **finely ground**, preferably passing through a sieve with 0.25-mm mesh.

Note: quicklime in lumps or granules can only be used as **a lime milk** for the disinfection of drained ponds (see Section 46).

BEWARE: Quicklime, hydrated lime and concentrated lime/water mixes can cause serious chemical burns. Avoid contact with skin and eyes as explained earlier in Section 46. If you do accidentally come in contact, wash immediately with plenty of water.

Making lime yourself

4. **Natural deposits rich in calcium carbonate**, such as limestone, shell deposits, or coral, can often be located near your farm. The best materials are usually white to light brown in colour. You should enquire about availability, for example at the Agricultural Department or the nearest office responsible for agricultural development. If this carbonate material is available, you can easily make lime yourself by remembering that:

- $CaCO_3$ + heat = CaO + CO_2 (gas in the air);
- CaO + water = $Ca(OH)_2$.

Note: if the material has too much clay, it will produce lime that sets and hardens when in contact with water and it is poorer for pond liming. 155

TABLE 12

Common types of liming materials

Basic chemical	Common name	Toxicity for fish	Relative price	Effectiveness[1]	Preferred when
Calcium carbonate $CaCO_3$	• limestone (90-95% $CaCO_3$) • dolomite (double carbonate of calcium/magnesium) • marl (20-80% $CaCO_3$) • others: basic slag, coral shells, etc.	Low	Low	Low and slow NV 100	• water pH above 4.5 • fish are present
Calcium hydroxide $Ca(OH)_2$	• hydrated lime, caustic lime slaked lime (aprox. 70% CaO)	Medium	Medium	Medium 0.7 kg = 1 kg $CaCO_3$ NV 136	• water pH below 4.5 • no fish present • for pest control (**Table 9**)
Calcium oxide CaO	• quicklime, unslaked lime, or burned lime	High	High	High and fast 0.55 kg = 1 kg $CaCO_3$ NV 179	• drained pond • for pest control (**Table 9**)

[1] NV = neutralizing value of the pure salts, in percent, with reference to $CaCO_3$ (NV = 100 percent)

5. **To prepare quicklime** from calcium carbonate material, proceed as follows.

(a) Mark out a circle 7 m in diameter on the ground. At its centre, mark a smaller circle 1 to 1.5 m in diameter. Draw two perpendicular lines through the centre of these two circles.

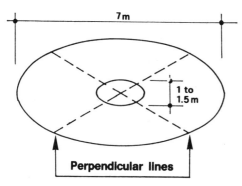

Perpendicular lines

(b) Stack four piles of logs about 1 m high in the outer circle leaving free spaces between the piles as shown in the illustration.
(c) Fill the free spaces separating the log piles with dried leaves, twigs and small branches, and cover these spaces with a layer of smaller logs.
(d) Break CaCO₃ material into pieces 10 to 15 cm in size. Place the broken pieces on top of the log pile to a depth of 50 to 90 cm.
(e) On a windy day, set the log pile on fire.
When all the wood is burned and while the material is still hot, sprinkle some water on it.
(f) Cover the burned pile with wet sacks, banana leaves or similar. Let it cool for at least 24 hours.
(g) Remove any over-burned or under-burned pieces. From 1 m³ of raw material, about 0.4 to 0.5 m³ of lime can be produced.

Note: quicklime should be stored in dry, airtight conditions; it will keep for several months in enclosed heaps or sealed bags. If it is exposed to water and air it returns to CaCO₃.

Preparing quicklime from $CaCO_3$

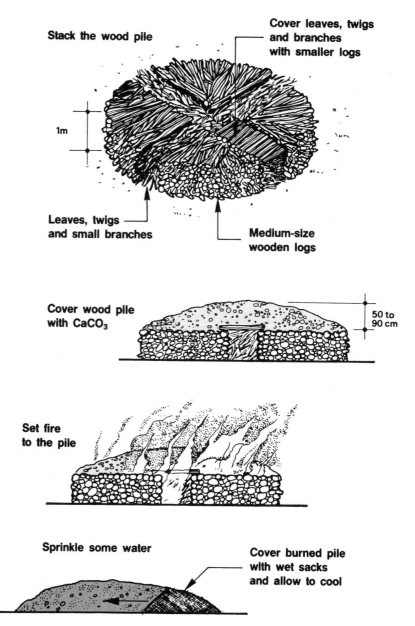

Stack the wood pile

Cover leaves, twigs and branches with smaller logs

Leaves, twigs and small branches

Medium-size wooden logs

Cover wood pile with $CaCO_3$

50 to 90 cm

Set fire to the pile

Sprinkle some water

Cover burned pile with wet sacks and allow to cool

157

6. **To prepare hydrated or slaked lime** from quicklime lumps or granules, proceed as follows.

(a) If necessary, break the lime lumps into 5- to 8-cm pieces.
(b) On clean, non-porous ground, build a mound 20 cm deep.
(c) Sprinkle it with water using a spray, at the ratio of about 12 l of water to 100 kg of lime. The reaction produces heat and some noise as the material breaks up.
(d) Turn over the heaped material while applying water, until the lime breaks up into a fine, white powder.
(e) The process is finished when no further heat is produced when adding more water.
(f) The slaked lime can then be screened for the required particle size.

Preparing slaked lime on the ground

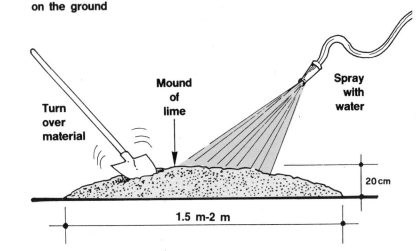

Turn over material

Mound of lime

Spray with water

20 cm

1.5 m-2 m

7. Slaked lime can also be prepared in a tank or drum. Use about three parts of water to one part of quicklime, add the lime to the water and let it stand for 24 hours.

Note: use this slaked lime as fresh as possible; if necessary, store it dry, in closed bags or sacks.

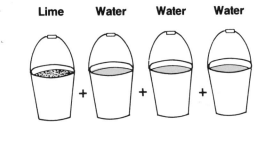

Lime   Water   Water   Water

+   +   +

1 part lime to 3 parts water

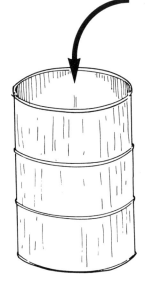

Preparing slaked lime in a drum

53 How to calculate the amount of lime needed

1. **The lime requirement** of ponds is defined as the amount of liming material needed:

- to neutralize the acidity of the pond bottom soil; and
- to increase the total alkalinity of the water above at least 25 mg/l $CaCO_3$.

2. The lime requirement therefore varies, depending on:

- **the nature of the bottom soil**, heavy clay requiring more lime than sandy soil;
- **the soil pH**, acidic soil requiring more lime than neutral soil;
- **the water total alkalinity**, soft water with a low TA requiring more lime than harder water with a higher TA.

3. It also varies with **the thickness of the bottom mud**. If the mud is 30 to 40 cm thick, the pond requires much more liming than if the mud is 5 to 10 cm thick.

Determining the lime requirement of a pond

Usually, **new ponds** require more liming than older ponds that have been regularly treated, for example once a year. There are therefore **several different treatments at different doses**, depending on the circumstances. Among these are:

- initial treatment of a new pond;
- routine treatment of a drained pond bottom;
- routine treatment of the pond water.

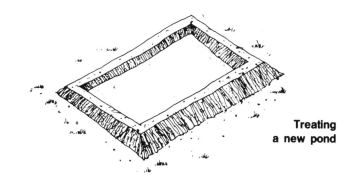

**Treating
a new pond**

5. **Initial treatment of a new pond**:

- according to the nature of the soil, spread from 2 000 kg/ha $CaCO_3$ (sandy soil) to 4 000 kg/ha $CaCO_3$ (heavy clay soil) on the dry pond bottom or an equivalent amount of another type of lime (see note below);
- fill the pond with water;
- one to two months later, determine the **total alkalinity of the water**; if it is greater than 25 mg/l $CaCO_3$, no more liming is required;
- if the TA is less than 25 mg/l $CaCO_3$, apply another dose of 2 000 kg/ha $CaCO_3$ to the water;
- one month later, determine the TA of the water again, and if it is greater than 25 mg/l $CaCO_3$, no more liming is required;
- if the TA is still less than 25 mg/l $CaCO_3$, apply a third dose of 2 000 kg/ha $CaCO_3$ to the water;
- check the TA one month later when it will normally be greater than 25 mg/l $CaCO_3$.

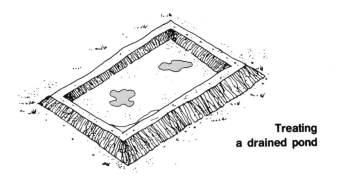

**Treating
a drained pond**

6. **Routine treatment of a drained pond bottom**: about once a year, apply one-quarter of **the total quantity of liming** material you required for the complete new pond treatment, as described above.

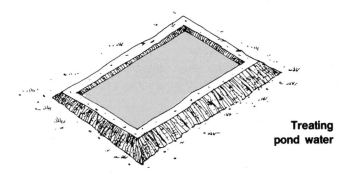

**Treating
pond water**

7. **Routine treatment of the pond water**: about once a month, check **the pH of the pond water at the end of the day**, and:

- **if the pH is less than 6.5**, add 150 to 200 kg/ha $CaCO_3$, checking the pH one week later and repeating liming if the pH is still too low;
- **if the pH is between 6.5 and 8.5**, check the TA of the water and if it is less than 75 mg/l $CaCO_3$, you could use lime to increase its value by adding one or several doses of 150 to 200 kg/ha $CaCO_3$ at weekly intervals until the TA is improved;
- **if the pH is greater than 8.5**, no liming is required.

Note: **do not forget** to correct the above quantities according to **the percentage of $CaCO_3$ present in the particular liming material** you are using. For example, if you are using a limestone containing 90 percent $CaCO_3$, multiply the amount of $CaCO_3$ suggested above by $100 \div 90 = 1.11$.

8. **If you do not use a calcium carbonate ($CaCO_3$) material for liming**, for example if you prefer to apply quicklime or hydrated lime on the drained bottom to control pests more efficiently (see Section 46), convert the above quantities according to the following equivalences:

$$100 \text{ kg } CaCO_3 = 70 \text{ kg } Ca(OH)_2 = 55 \text{ kg } CaO$$

Example

To lime your ponds, you require 2 000 kg/ha (or 20 × 100 kg/ha) $CaCO_3$. Instead, you could use either:

- $Ca(OH)_2$ at the rate of 20 × 70 kg/ha = 1 400 kg/ha; or
- CaO at the rate of 20 × 55 kg/ha = 1 100 kg/ha.

Note: when applied directly in water, the effects of quicklime or hydrated lime are much more rapid and need to be carefully monitored. In such cases it may be useful to apply some of the lime as $CaCO_3$ to moderate the speed or effects of the other materials.

Introduction

1. Usually liming materials and fertilizers are applied separately. **Liming then should be done at least two weeks, preferably one month, before any application of fertilizers** (see Section 61).

2. Annual liming therefore is carried out in different periods of the year according to the pond management schedule.

a) In **temperate climates**, such as in Europe, **fattening ponds** may be limed in the autumn just after they have been drained and harvested. Quicklime or hydrated lime is spread on the wet pond bottom, and fertilizers are applied in spring only. On the other hand, **nursery ponds** are limed in the spring just before they are half-filled with water, fertilizers being distributed later.

b) In **tropical climates**, ponds are best limed as soon as the fish have been harvested and at least two weeks before new batches of fish are stocked. Fertilizers are then applied 15 to 30 days after liming.

Liming drained ponds

3. Drained ponds can easily be limed by distributing the **finely ground liming material** equally over the whole of the bottom surface. If the objective of liming does not include pest control (see Section 46), **the bottom surface should preferably be dry**.

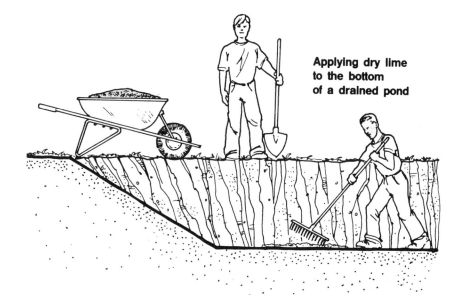

Applying dry lime to the bottom of a drained pond

4. It will be easier to distribute the liming material evenly if **it is first diluted in some water** before being thrown into the pond. Buckets, wheelbarrows or old steel drums can be used to dilute the lime.

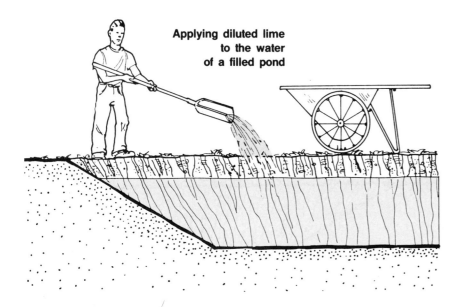

Applying diluted lime to the water of a filled pond

Apply diluted lime from the banks in a small pond

Apply diluted lime from a boat in a larger pond

5. Lime diluted in water can be applied using a large wooden spoon. In small ponds this can be done directly from the banks, but if the pond is large it may be necessary to use a boat or a floating platform.

Note: be particularly careful **if you use quicklime in the presence of fish**. In any one day, **do not apply** more than 200 kg/ha CaO, and frequently check the water pH at the end of the day. Be sure that it always remains below 9.5, or your fish might die.

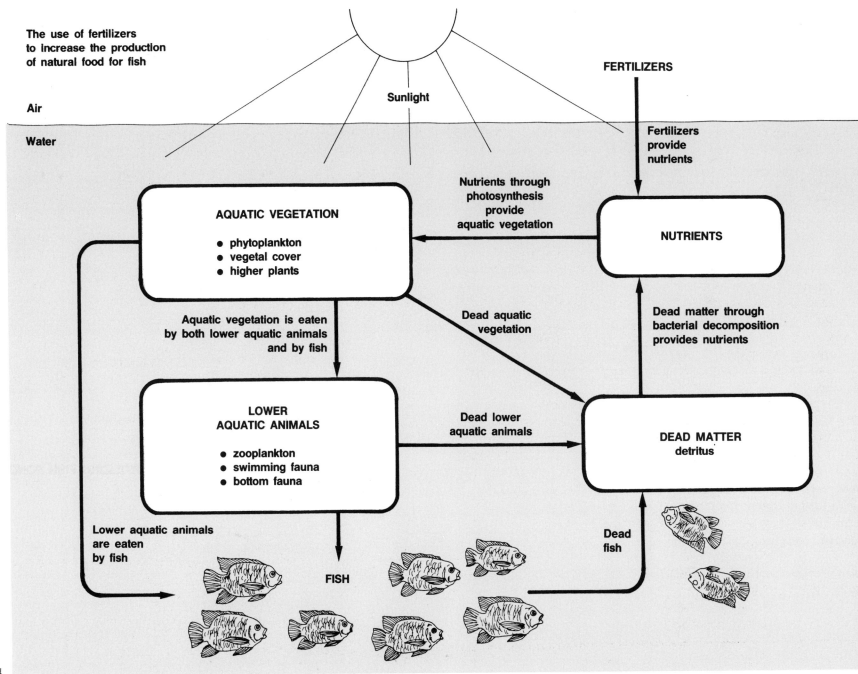

The use of fertilizers to increase the production of natural food for fish

FERTILIZERS

Sunlight

Air

Water

Fertilizers provide nutrients

Nutrients through photosynthesis provide aquatic vegetation

AQUATIC VEGETATION

- phytoplankton
- vegetal cover
- higher plants

NUTRIENTS

Aquatic vegetation is eaten by both lower aquatic animals and by fish

Dead aquatic vegetation

Dead matter through bacterial decomposition provides nutrients

LOWER AQUATIC ANIMALS

- zooplankton
- swimming fauna
- bottom fauna

Dead lower aquatic animals

DEAD MATTER
detritus

Lower aquatic animals are eaten by fish

Dead fish

FISH

Introduction

Fertilizers are natural or synthetic substances that are used in ponds increase **the production of the natural food organisms** to be eaten by e fish. These organisms include phytoplankton, zooplankton and sects (see Chapter 10, **Management, 21/2**). They are all part of a mplex **food web*** converging toward fish production. By increasing the ailability of major nutrients, **fertilizers promote the development of anktonic algae, which provide food for many fish**. Fertilization also ads to the development of animals which feed on algae, including me fish such as the Chinese silver carp and the Nile tilapia.

When a fertilizer is added to a fish pond, **the chemicals** it contains ssolve in the water, where:

- a portion is usually rapidly **taken up by the phytoplankton** present, either **to be stored**, sometimes in quite large proportions, or **to be assimilated** and used for growth, reproduction, etc.;
- another portion is **attracted by and becomes attached to** the organic and mineral particles present, both in the pond water (see Section 23) and in the upper layers of the bottom mud or soil.

This second portion may also assist the development of **bacteria**, ponsible for the decomposition of organic matter. The decomposition organic matter may in turn release more nutrients back into the mud water. The chemicals attached to soil particles may also later be eased back into the water slowly, over a long period of time. They y also migrate deeper into mud and soil, where they will no longer ct the water body, unless the pond bottom is dried or ploughed (see tion 25).

Most of these phenomena are linked with and controlled by **water lity** and in particular temperature, pH, alkalinity and dissolved oxygen el.

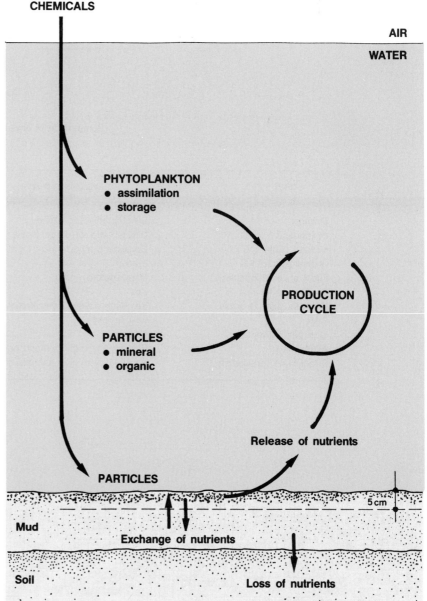

The action of chemicals contained in fertilizers when dissolved in water

TABLE 13

Comparison of organic and inorganic fertilizers

Item	Organic fertilizers	Inorganic fertilizers
Storage	Difficult, only short time	Easy, possibly for long time
Distribution	Difficult, esp. on larger scale	Easy
Mineral content	Consistent, high to very high	Variable, low
Organic matter	Present	Absent
Effect on soil structure	Improvement	No
Direct food for fish	Yes	No
Decomposition process	Yes, with oxygen consumption	No
Price	Low to medium	High to very high
Cost per nutrient unit	Higher	Lower
Availability	Possibly in neighbourhood or even on own farm	Commercial suppliers only; sometimes imported
Direct pond fertilization	Possible by raising animals on or near the pond	Not feasible

Pond fertilizers form two distinct groups:

- **mineral or inorganic fertilizers**, which contain only mineral nutrients and no organic matter; they are manufactured industrially to be used in agriculture for improving crop production and they can be obtained from specialized suppliers;
- **organic fertilizers**, which contain a mixture of organic matter and mineral nutrients; they are produced locally, for example as wastes from farm animals or as agricultural wastes.

Both types of fertilizer have **advantages** and **disadvantages**, as listed Table 13. Select the most appropriate type of fertilization for your own eeds. Selection may depend not only on local availability but also on e size of the farm.

) **Small-scale fish farms** tend to rely on organic fertilizers as they are cheap and available locally.

) **Large-scale fish farms** most often adopt inorganic fertilizers, as they are more easily stored and distributed.

The best results can often be obtained with **the combined use of both es of fertilizer**.

aking the best use of fertilizers

When using fertilizers **to increase fish production in your ponds, you ould aim to establish and maintain a dense growth of planktonic algae** ytoplankton) and **zooplankton**, which should colour the water a rich ade of green. Such dense planktonic growth is often called **a plankton om**.

To establish and maintain a good plankton bloom at minimum cost, tch for the following points.

Pond water and bottom soil should be neutral or slightly alkaline. Lime them if necessary (see Chapter 5).

(b) If present, **bottom mud** should be good quality, not too thick and mostly made of fine detritus; too much **cellulose*** slows down its decomposition. Control the emersed vegetation and the mud thickness, if possible by draining and drying (see Section 25).

(c) **Reduce the competition for nutrients and sunlight** by controlling the floating and submersed vegetation (see Section 49).

(d) **Reduce the water exchange rate** as much as possible to avoid draining away water rich in nutrients and plankton.

(e) **Fertilize each pond** according to its particular characteristics; for example, use more fertilizer:

- if the pond is new, and good mud has not yet formed;
- if the water supply is poor in nutrients (see the chart below);
- if the bottom soil is sandy rather than clayey.

(f) **Add more fertilizer** as needed according to the plankton density (see Section 101, **Management, 21/2**), using small amounts regularly, if possible.

Chemical qualities of a poor water supply

Phosphates	Less than 0.1 mg/l
Nitrates	Less than 2 mg/l
Potassium	Less than 0.1 mg/l
Calcium and magnesium	Less than 15 mg/l

10. If supplies of fertilizer are limited, give priority to the ponds where the availability of **natural food is most important**, for example nursery ponds and broodstock ponds.

11. **Do not fertilize a pond** if:

- fish production does not depend on the use of **natural food**;
- **the exchange rate of the water** is excessive;
- there is too much **emersed or other aquatic vegetation**;
- the water is too muddy or dark-coloured, and **transparency is limited**;
- the **plankton turbidity** is too high.

167

Deciding about the need for fertilizers

12. **The Secchi disc transparency** (see Section 23) can be used as a simple method for judging plankton turbidity and the need for additional fertilization of a fish pond. Depending on the value observed, control and manage the pond as shown in the following chart.

Fertilization needs

Secchi disc transparency	Management/control

Less than 25 cm	No fertilization:
	    Closely observe fish for signs of dissolved
	        oxygen depletion (see Section 25)
	    Increase water inflow, if necessary

25–40 cm	No fertilization: Regularly observe fish behaviour
40–60 cm	Routine fertilization necessary
More than 60 cm	Routine fertilization necessary, possibly with an increased dose

13. If you do not have a Secchi disc, you can use your arm instead. Stick your arm vertically underwater. As long as **your hand is not visible when your elbow is at the water surface**, there is no need for fertilization.

Note: **avoid over-fertilization**. It is both wasteful and dangerous for your fish.

14. The next sections tell you more about organic and inorganic fertilizers, and how best to use them.

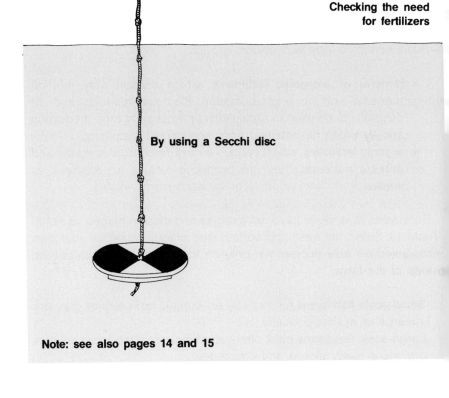

By using a Secchi disc

Note: see also pages 14 and 15

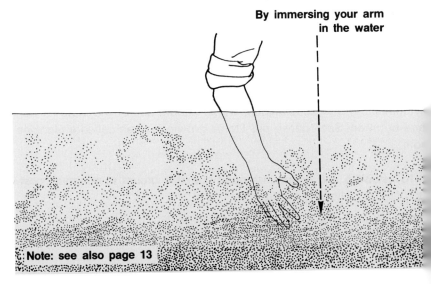

By immersing your arm in the water

Note: see also page 13

Inorganic fertilizers in fish farming

erent kinds of inorganic fertilizer

An agricultural inorganic fertilizer may contain several types of
·ients:

- **primary nutrients**: nitrogen (N), phosphorus (P) and potassium
  (K);
- **secondary nutrients**: calcium, magnesium and sulphur, for example;
- **trace nutrients** in very small quantities, such as manganese,
  zinc, copper and iron.

However, the fertilizers are named only according to **the primary
·ients** they contain.

Fertilizers which contain only **one or sometimes two primary
nutrients** retain their chemical name such as superphosphate (P) or
ammonium phosphate (N + P) (see **Table 14**).
Those with **two or three primary nutrients**, known as **mixed fertilizers**, are referred to by their **NPK grade**, that is, their percentages,
as weight, of the three primary nutrients;

- **nitrogen N**, as pure nitrogen;
- **phosphorus P**, as the equivalent in phosphoric acid P_2O_5;
- **potassium K**, as the equivalent in potash K_2O.

Example

100 kg of a mixed fertilizer called **8-8-2** contain 8 kg nitrogen, 8 kg P_2O_5
equivalent and 2 kg K_2O equivalent, the remaining weight being mostly
made up of **inactive filler**. Similarly, a **10-20-0** fertilizer contains
10 percent N, 20 percent P_2O_5 equivalent and 0 percent K_2O equivalent.

·: to determine how much pure phosphorus P or pure potassium K a
·d fertilizer contains, multiply the equivalent values as follows:

- $P = P_2O_5 \times 0.44$;
- $K = K_2O \times 0.83$.

Example

The 100 kg of **8-8-2** mixed fertilizer (see previous example) contain:

- 8 kg $P_2O_5 \times 0.44 = 3.52$ kg phosphorus;
- 8 kg $K_2O \times 0.83 = 6.64$ kg potassium.

3. **Inorganic fertilizers** most commonly used in fish farming are listed in
Table 14. Exact concentrations of nutrients may vary from one supplier to
another, according to the origin of the fertilizer. In addition, many types of
mixed fertilizer are commercially available according to the local needs
of plant crops. Contact your local office for agricultural development to
obtain information about locally available inorganic fertilizers.

Selecting inorganic fertilizers

4. Generally, **phosphorus is the limiting primary nutrient most often
missing in natural water supplies** for good growth of planktonic algae.
Therefore, **phosphate fertilizers** are usually the most effective inorganic
fertilizers for fish ponds in most regions of the world.

5. **Nitrogen fertilizers** are sometimes useful, especially **in temperate
climates** and **in watersheds where agriculture is little developed**. They
are mostly used to avoid an unbalanced P:N ratio in the water, which is a
risk when using phosphate fertilizers alone. If the P:N ratio is too high,
blue-green algae may bloom instead of the more desirable green algae.
In the tropics, where fixation of nitrogen by bacteria and algae is much
more active, nitrogen fertilizers are less needed. **In older ponds** with a
good layer of mud, they are usually unnecessary.

Note: some nitrogen fertilizers, such as ammonium compounds and
urea, are **acid forming**. When applied to a pond, additional liming may be
required to maintain water pH and total alkalinity within adequate limits
(see Section 51).

TABLE 14

**Concentration of primary nutrients in common inorganic fertilizers
(in percentage, by weight)**

Fertilizer	Phosphorus		Nitrogen	Potassium		Calcium present	Water solubility
	eq. P_2O_5	P[1]	N	eq. K_2O	K[2]		
PHOSPHATE							
Basic slag	16-20	7.0-8.8	—	—	—	40% CaO	Poor if Ca high
Superphosphate	14-20	6.2-8.8	—	—	—	Yes	High (85%)
Triple superphosphate	44-54	19.4-23.8	—	—	—	Yes	High (85%)
NITROGEN							
Ammonium nitrate	—	—	33-35	—	—	—	High
Ammonium sulphate	—	—	20-22	—	—	—	High
Ammonium phosphate	20-48	8.8-21.1	11-16	—	—	—	High
Di-ammonium phosphate	48-52	21.1-22.9	18-21	—	—	—	High
Calcium nitrate	—	—	15-16	—	—	Yes	High
Sodium nitrate	—	—	15-16	—	—	—	High
Urea	—	—	42-47	—	—	—	High
POTASSIUM							
Kainite ($MgSO_4 + KCl$)	—	—	–	20	16.6	—	High
Potassium nitrate	—	—	13-14	44-46	36.5-38.2	—	High
Potassium sulphate	—	—	—	45-54	37.4-44.8	—	High
Muriate of potash	—	—	—	50-62	41.5-51.5	—	High

[1] Phosphorus: to be multiplied by 2.29 to obtain P_2O_5 equivalence
[2] Potassium: to be multiplied by 1.2 to obtain K_2O equivalence

TABLE 15

Criteria for the use of inorganic fertilizers

	Phosphate fertilizers	Nitrogen fertilizers	Potassium fertilizers
Water quality: desirable concentrations for good algal production:	Phosphates > 0.2 mg/l Total P > 0.4 mg/l	Nitrates > 2 mg/l Total N >1.5-3 mg/l	Potassium > 1 mg/l —
Best P to N ratio	Enough total P present for total N to be used: P to N = 1:4 to 1:8		—
Inorganic fertilizer advisable for:	• poor water/soil • acid soil or light soil and/or water poor in Ca: prefer basic slag • soil/water richer in Ca and/or heavy soil: prefer superphosphate	• poor water/soil • new ponds • ponds with no bottom mud • nursery ponds • more intensive cultural system, at higher fish density	• poor water/soil • water total alkalinity less than 25 mg/l $CaCO_3$ • ponds in swampy areas • peaty soil • hard pond bottom and little aquatic vegetation
Typical amount, per hectare, for one production cycle	• 30-60 kg eq. P_2O_5	• 40-100 kg N (check P:N ratio)	• 35 kg K_2O to 60-80 kg in peaty soil

6. **Potassium fertilizers** are not generally necessary except in specific locations where there is a deficiency in potassium. This situation may happen in ponds built in **swampy areas** or in **peaty soils**. Additional potassium may also be useful in ponds with **a hard bottom** and little aquatic vegetation.

7. Before spending too much money on inorganic fertilizers, you should check on the following.

(a) **The chemical quality of the water supply**: have at least one good chemical analysis carried out, checking on total phosphorus, phosphates, nitrogen, nitrates and potassium concentrations as well as total alkalinity and pH. If possible, check again during different seasons.
(b) **The nature of the pond bottom soil** (sandy/light or clayey/heavier) and its **chemical characteristics** should be examined, such as pH and concentrations of calcium and primary nutrients.
(c) **The water solubility of the fertilizer** is important (see **Table 14**), especially for phosphate fertilizers. Look for:

- **a highly soluble chemical**, unless you want a longer-term effect, for example correcting the water quality at the same time;
- **a small particle size**.

8. Further assistance is given in **Table 15**. Also remember that some of the substances you are using for pest control (see Sections 46 and 47) have a fertilizing effect. Calcium cyanamide, for example, contains 18 to 22 percent nitrogen.

Mixing inorganic fertilizers with other substances

9. To save on time and labour, inorganic fertilizers are commonly mixed with each other or with substances such as organic fertilizers or liming materials.

10. However, you should **avoid certain mixtures** in which the components react with each other, and fertilizer quality deteriorates.

11. For maximum safety, check the chart on the opposite page and apply the following rules:

(a) **Never mix** the following **together** and wait for at least two weeks between separate distributions:

- phosphate fertilizers and liming materials;
- nitrogen fertilizers and liming materials;
- basic slag and organic fertilizers.

(b) You may **mix together, but only just before use**:

- potassium fertilizers and liming materials or basic slag;
- phosphate and nitrogen fertilizers;
- potassium fertilizers and superphosphate, nitrogen or organic fertilizers.

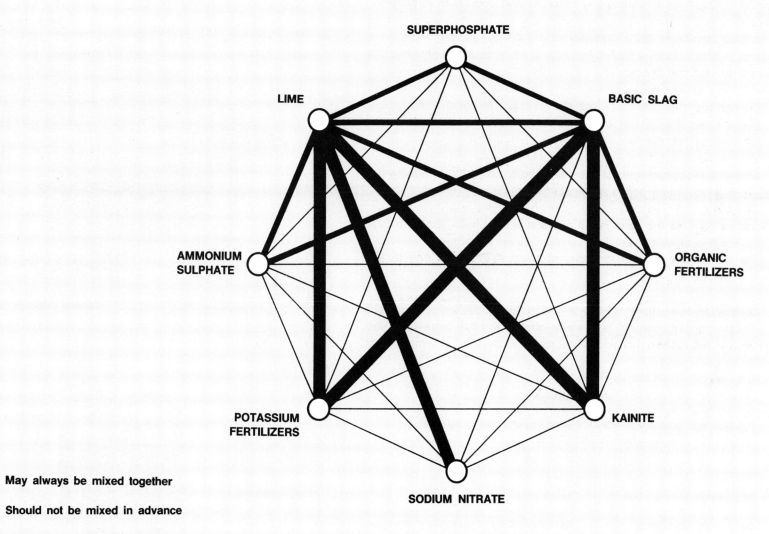

SUPERPHOSPHATE

BASIC SLAG

LIME

ORGANIC
FERTILIZERS

AMMONIUM
SULPHATE

KAINITE

POTASSIUM
FERTILIZERS

SODIUM NITRATE

———————— May always be mixed together

━━━━━━━━ Should not be mixed in advance

▬▬▬▬▬▬▬ Should never be mixed together

173

Storing inorganic fertilizers

12. It is best to avoid having to store inorganic fertilizers for too long. Buy in small quantities, only as much as you need, and store them for the shortest possible time, **particularly nitrogen and potassium fertilizers**.

13. Protect fertilizers from humidity and rain by storing them on a wooden platform under a roof, by wrapping them in plastic sheets or by doing both of these things.

14. **Do not buy** any fertilizer which shows signs of being wet.

Fertilizing your fish ponds

15. Unless the production cycle is very short, inorganic fertilizers are usually applied to fish ponds at regular intervals:

- **as soon as the pond is full of water** and at least ten to 15 days before stocking with fish;
- **during the production cycle**, at short intervals of one to two weeks.

16. Application of inorganic fertilizers during the production cycle should be based on observations of both **water quality** and **fish behaviour** (see Section 60). The following chart will assist you in deciding whether or not to fertilize your pond.

Protecting stored fertilizers

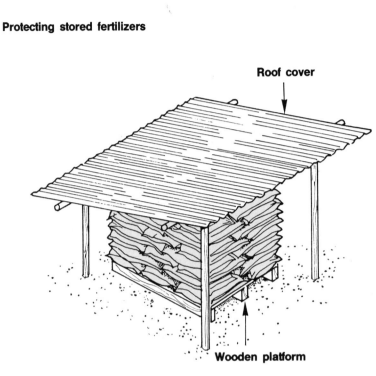

Roof cover

Wooden platform

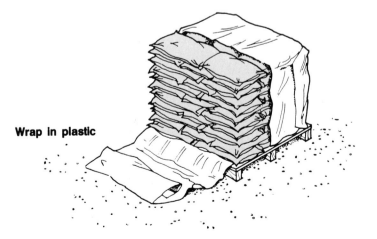

Wrap in plastic

Mixing fertilizers

Fertilize your pond inorganically if at least one of the following is present

	Yes	No
Water temperature at noon	Above 16°C	Below 16°C
Secchi disc transparency	More than 40 cm	Less than 40 cm
Water pH at sunset	Less than 9	More than 9
Dissolved oxygen before sunrise	More than 3 mg/l	Less than 3 mg/l

Choosing the quantity of inorganic fertilizer

17. As you learned earlier, the kind and amount of inorganic fertilizer to use can vary greatly from one pond to another. It is therefore not possible to recommend any specific mixture or dosage which would give best results in all locations. **Table 15**, however, summarizes points to consider for further guidance.

18. It is best **to determine the amount of fertilizer needed in each of your ponds** by trial and error as follows:

(a) When the pond is full of water, make **an initial application** of phosphate fertilizer equivalent to 125 to 175 g P_2O_5/100 m^2 or 12.5 to 17.5 kg/ha.

(b) **If additional nitrogen** is required, apply sufficient nitrogen fertilizer to give a ratio of 4 to 8 N for each phosphorus unit, taking into account if possible the nitrogen already present in the pond to avoid wasting fertilizer. The richer the water and the higher the fish density, the higher this nitrogen proportion should be.

Example

You want to plan a fertilization programme for your pond. You obtain superphosphate containing 20 percent P_2O_5 equivalent (**Table 14**) and decide to apply it at the rate of 150 g P_2O_5/100 m^2.

(a) How much **superphosphate** do you need per 100 m^2?

● Each 100 g fertilizer provides 20 g P_2O_5, so to provide 150 g you need (100 g × 150 g) ÷ 20 g = 750 g superphosphate.

(b) How much **nitrogen** do you need per 100 m^2 for a ratio P:N = 1:6?

● 150 g P_2O_5 × 0.44 = 66 g **phosphorus** (P);
● 66 g P × 6 = approximately 400 g nitrogen.

(c) How much **nitrogen fertilizer** should you apply per 100 m^2?

If there is little nitrogen already in the pond or if you are not sure of the levels, add nitrogen **according to the phosphorus and nitrogen amounts** calculated above. You will need 400 g N. If, for example, you use ammonium nitrate with 35 percent N (**Table 14**), you need to apply (100 g × 400 g) ÷ 35 g = approximately 1 140 g/100 m^2 pond.

If you know what the nitrogen level is (for example, by measuring it at the time), you may save on fertilizer. If, for example, total nitrogen N = 3 mg/l = 3 g/m^3 and water is 1 m deep:

● each 100 m^2 area of pond contains 100 m^2 × 1 m = 100 m^3 of water;
● nitrogen present per 100 m^2 of pond = 100 m^3 × 3 g/m^3 = 300 g;
● as you need 400 g, you need only to add 100 g nitrogen/100 m^2;
● if you use ammonium nitrate as above, you need about (100 g × 100 g) ÷ 35 g = 286 g/100 m^2 pond.

(c) Seven to ten days after applying the first fertilizer dose, measure the **Secchi disc transparency** before applying a new dose, and decrease or increase the previous amounts of fertilizer accordingly.

(d) Repeat this process at regular intervals of seven to 15 days so as to maintain the **SD transparency** at between 40 and 60 cm throughout the production cycle.

(e) Keep checking on **water quality** and **fish behaviour** to further modify your fertilization programme if necessary (see criteria in chart on page 174).

Note: if **potassium fertilizer** is required, use 0.35 to 0.80 kg K_2O/100 m^2 according to local conditions (see **Table 15**).

Distributing inorganic fertilizers

19. To apply inorganic fertilizers, remember the **following recommendations**.

(a) Apply regularly at **very short intervals** of preferably seven to 15 days, especially if the pond bottom is sandy and little mud has built up over it.

(b) For each treatment, use **small doses** of fertilizer.

(c) Before the treatment and for a few days after, **reduce the water inflow** as much as possible.

(d) For best results, **never throw solid fertilizers directly into the pond water**. This is especially important for **phosphate fertilizers**, because the bottom mud or soil can quickly turn the soluble phosphorus into insoluble compounds, which are then of limited use for the pond water.

20. There are **two main methods** for fertilizing ponds.

- using dry, inorganic fertilizers; and
- using dissolved inorganic fertilizers.

21. **You may use the dry inorganic fertilizers directly** and let them dissolve slowly. Water currents help to disperse the dissolved chemicals through the entire pond area. In **small ponds**, use at least one fertilizing point per 1 000 m² of water area. In **larger ponds**, use two to three points per hectare.

22. A number of methods for distributing dry inorganic fertilizers are shown on this page and page 177.

(a) From a wooden post, suspend **a small bag** made of cotton or burlap, about 30 cm underwater. In this bag enclose the seven- to 15-day dosage of fertilizer for the water area concerned. At the end of this period, empty the insoluble filler from the bag and add a new dosage of fertilizer. You could also use a perforated can or a basket.

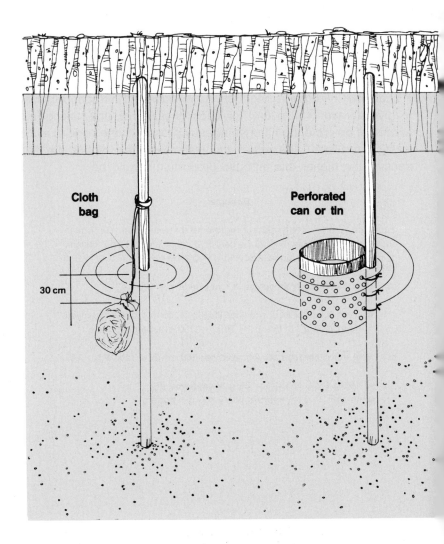

Cloth bag

Perforated can or tin

30 cm

b) Use **a floating perforated container** such as a woven basket or a plastic can with holes, attached to the inside of a car tyre inner-tube. Use this container as you were told in item (a) above.

(c) Submerge **a wooden platform** 30 cm underwater and set it at least 30 cm above the pond bottom. The doses of fertilizer are placed on the platform, either directly, if there is no risk of it being swept away or, when water currents are too strong, in open fertilizer bags.

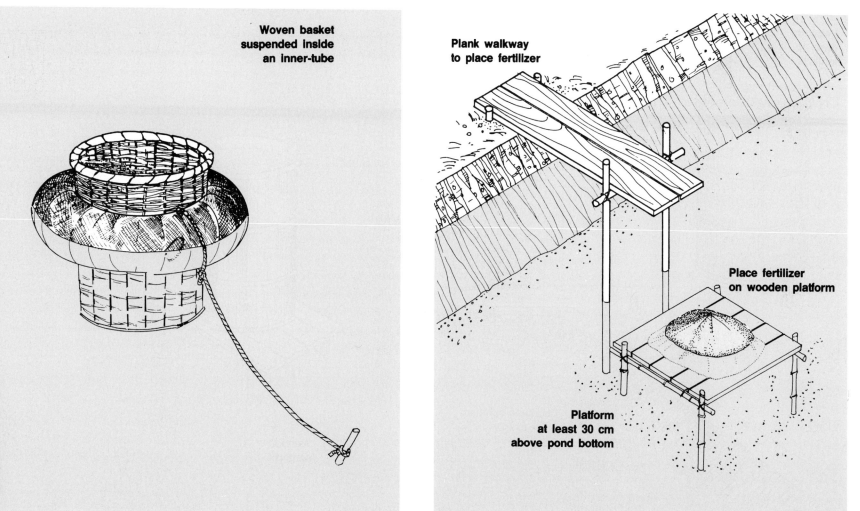

Woven basket
suspended inside
an inner-tube

Plank walkway
to place fertilizer

Place fertilizer
on wooden platform

Platform
at least 30 cm
above pond bottom

23. For better results **you may use dissolved inorganic fertilizers** in two ways.

(a) **For periodic pond fertilization**, dissolve the required dose of fertilizer well in a large container such as a clean 200-l metal drum filled with pond water. Using buckets, distribute this solution evenly over the whole pond surface from the banks or, if necessary, from a boat.

(b) **For continuous pond fertilization**, build one or more emersed wooden platforms in the pond. On these platforms, permanently install a large container such as a 200-l metal drum, equipped with a small outflow valve about 10 cm above its base. You will need about five such containers per hectare. Fill each container with 100 to 150 l of water and dissolve the appropriate dose of fertilizer in it. Open the valve just enough to let the solution flow out steadily over a period of several days.

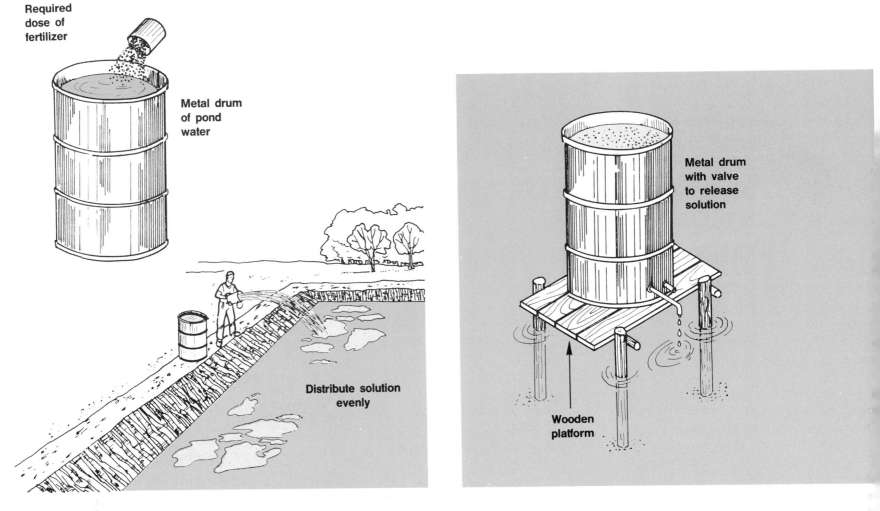

Required dose of fertilizer

Metal drum of pond water

Distribute solution evenly

Metal drum with valve to release solution

Wooden platform

62 Organic fertilizers: animal manures

1. In many instances, especially for small farmers, organic fertilizers are the most effective way of increasing natural food supply in ponds to improve fish production.

Different kinds of organic fertilizer

2. Several kinds of organic material, mostly waste materials, can be used as organic fertilizers. Most common are the following:

- animal manures, mostly from farm animals;
- slaughterhouse wastes;
- agro-industrial wastes;
- biogas slurry;
- cassava fermentation;
- natural vegetation;
- compost, a mixture of various kinds of organic matter.

3. In the next paragraphs, you will learn about **animal manures**. You will learn more about the other kinds of organic fertilizer in Sections 63 and 64.

Animal manures as organic fertilizers

4. As pond fertilizers, **animal manures** have such great advantages that they should be preferred whenever possible.

5. As **direct food**, they can partly replace supplementary feeds (see Section 103, **Management, 21/2**). For example, on manuring days additional feeding may be cancelled. Some fish, such as the Nile tilapia, may even be produced in large quantities without any additional feeding.

6. They are a source of additional **carbon dioxide** (CO_2), which is very important for the efficient utilization of the nutrients present in the water. This is especially so when used together with inorganic fertilizers.

7. They increase the abundance of **bacteria** in the water, which not only accelerate the decomposition of organic matter (see Section 20), but also serve as food for the **zooplankton**, which in turn also increases in abundance.

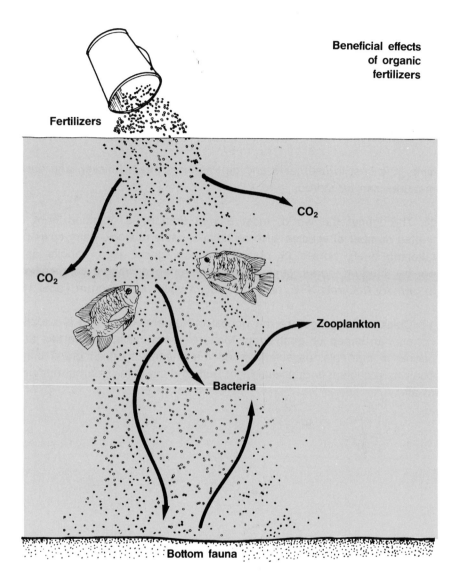

Beneficial effects of organic fertilizers

Fertilizers

CO_2

CO_2

Zooplankton

Bacteria

Bottom fauna

8. They have beneficial effects not only on the **bottom soil** structure but also on the **bottom fauna**, such as the chironomid larvae.

9. However, animal manures also have **some disadvantages**, mostly related to their low content in **primary nutrients**, their negative effect on **dissolved oxygen content** and the reluctance of some **fish farmers** to use animal wastes directly in fish ponds.

179

10. The **chemical composition** of organic manure varies greatly according to the animal from which it originates — namely the species, age, sex, the nature of its diet — and according to the way the manure is handled, namely its relative freshness, conditions of storage and rate of dilution with water. In some cases, total wastes made of dung and urine are available, while in others only solid wastes can be collected.

Note: it is best to use cattle and pig manures fresh. Chicken and duck manures can be stored.

11. Throughout the world, most animal manure is obtained from a **limited number of species** such as buffalo, cattle (bullock, dairy cows or fattening beef), horses or donkeys, sheep, goats, pigs, rabbits and poultry (chicken, ducks, geese). Examples of solid manure **composition in primary nutrients NPK**, on a dry weight basis, are given in **Table 16**.

12. **Chicken droppings** are the richest in nutrients. **Pig dung** is usually richer than **sheep or goat dung**. Manures from **cattle and horses** are poorer in nutrients, especially when the animals feed on grass only. Their fibre content is relatively high. **Buffalo dung** is the poorest manure of all.

13. The total amount of manure produced daily by various animals depends mainly on their **live weight**, as shown in **Table 17**. Pigs, for example, produce a daily average of about one-tenth of their live weight in **total wet wastes**, consisting of solid wastes and urine. A little less than half of this is made up of **solid manure**.

14. Depending on the conditions under which the manure can be collected, it may be mixed with other kinds of organic matter such as

- **litter material**, for example straw used for the bedding of horses and cattle or rice husks used in some chicken houses;
- **spilled animal feed**, for example in houses where poultry are raised intensively.

15. In these cases, the quantity of organic fertilizer collected daily may be higher. Its quality also changes according to the materials added to it.

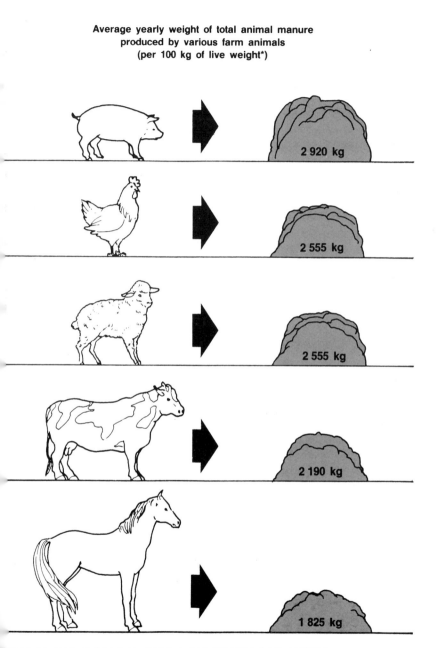

Average yearly weight of total animal manure produced by various farm animals (per 100 kg of live weight*)

2 920 kg

2 555 kg

2 555 kg

2 190 kg

1 825 kg

...ased on manure free from bedding and collected from confined ...mals

...e: see also **Table 17**

Selecting the best animal manures for your fish pond

16. If several types of manure are available, **choose the best for fertilizing your pond** according to the following criteria.

(a) The manure should be **easily soluble and dispersable in water**. Liquid manure or solid poultry wastes are preferred, because cow or horse dung usually contains a lot of insoluble **cellulose** especially if mixed with stable litter.

(b) It should be in **small particles** rather than in lumps.

(c) Use it **as fresh as possible**. Large losses of nitrogen and carbon occur during storage, especially if the manure is left in the open air and in the rain.

(d) Make sure it has a **high nutrient content**, as discussed above (see also **Table 16**).

(e) Manure should be **easy to collect**. Housed or corralled animals produce more concentrated manure than free-roaming ones. Animal housing can be designed to improve the collection and distribution of manure to the ponds (see Chapter 7).

Note: in **new ponds with sandy soil**, cattle manure with higher fibre content may be preferable to help form the bottom mud.

TABLE 16

Examples of the NPK composition of animal manures[1]
(percent of oven-dry weight)

Animal/poultry	Country	Nitrogen N	Phosphorous P	Potassium K
Buffalo dung	China	1.50	0.55	0.40
	India	0.75	0.20	2.00
Horse dung	India	1.88	0.52	1.00
	USA	2.00	1.20	0.80
Cattle dung	India	1.65	0.44	0.83
	UK	2.98	0.41	1.78
Sheep dung	India	1.55	0.70	0.72
	USA	1.89	1.35	0.54
Goat dung	(Asia)	2.04	0.73	0.47
Pig dung	China	2.66	1.37	1.47
	USA	3.03	1.66	1.60
Rabbit droppings	—	1.72	1.30	1.08
Goose droppings	Hungary	0.6	0.22	0.83
Duck droppings	Mean values	2.15	1.13	1.15
	Hungary	1.00	0.62	0.50
Chicken droppings	China	5.14	1.98	2.03
	India	2.87	1.28	1.95
	USA	4.59	2.33	1.96

[1] R.V. Misra and P.R. Hesse, *Comparative analyses of organic manures*, FAO/UNDP Regional Project RAS/75/004, Project Field Document 24, 97p.

TABLE 17

Estimated daily production of farm animal wastes

Animal/poultry	Live weight LW (kg)	Total wet wastes[1] per day % LW	kg	Solid wastes per day % LW	kg	Dry matter Total wet wastes[1] %	g/d	Solid wastes %	g/d	Total fresh wastes[1] (Solids only) Kg/100 kg LW/day	Total dry matter[1] (Solids only) Kg/100 kg LW/day
Buffalo	250	6.4	16	4.5	11	17	2 720	19	2 100		
	500	5.7	28	4.0	20		4 760		3 800		
Cattle	210	6.2	13	4.7	9		1 950		1 800	6.0 (4.5)	0.95 (0.85)
	350	6.0	21	4.3	15	15	3 150	20	3 000		
	450	6.0	27	4.2	19		4 050		3 800		
	550	6.2	34	4.4	24		5 100		4 800		
Horse	380	5.0	24	3.5	13	20	4 800	23	3 000	5.0 (3.5)	1.0 (0.7)
Sheep/goat	30	7.0	2.1	3.5	1.0	28	590	35	368	7.0 (3.5)	1.9 (1.2)
Pig	13-36	10.0	2.5	5.3	1.3		350		390	8.0 (4.3)	1.1 (0.9)
	36-54	8.0	3.0	4.3	1.6	14	420	20	480		
	54-72	6.0	3.5	3.2	1.9		490		570		
	72-90	4.5	3.5	2.7	1.9		490		570		
Duck	2-3	11	0.275	—	—	43	118	—	—	11.0	4.7
Chicken	1-1.5	7	0.080	—	—	45	36	—	—	7.0	3.2

[1] Solid wastes and urine

17. Fish ponds are usually fertilized with animal manure **at least ten to 15 days before stocking with fish**. In drained ponds, the manure is applied to the pond floor just before refilling with water (see paragraphs 30 to 33). In undrained ponds, the manure is applied to the water (see paragraphs 36 to 38).

18. After the first application, the pond should be fertilized **at regular intervals** throughout the fish production cycle. For best results, fertilize your ponds with manure **frequently**, at short intervals preferably not less than once a week. Daily applications are best.

19. As you have already learned (see Section 61), you should **monitor your pond carefully** during fertilization to avoid fish losses. This is especially important if you are using animal manure. Continue to fertilize a pond only if:

- **the water quality** remains acceptable (see the following chart);
- **the behaviour of the fish** remains normal (see Section 25).

20. If you do not have enough animal manure to fertilize all your fish ponds, **give priority to those** where its effects are especially beneficial:

- **new ponds**, particularly if their bottom soil is sandy and contains little organic matter;
- **nursery ponds**, if frequently drained and/or where limited time is available to boost up natural food production.

Recent research has shown that best fertilization results are obtained in ponds where **a constant flow of organic carbon** is maintained through the entire **food web** (see Section 101, **Management, 21/2**). Dense populations of planktonic algae, bacteria and zooplankton are thus established and kept relatively stable, preventing sudden high peaks of abundance of various groups. The water quality, such as dissolved oxygen content, also remains much more stable and favourable for fish production.

When to fertilize your pond organically

	Yes	*No*
Water temperature at noon	Above 20°C	Below 20°C
Secchi disc transparency	More than 40 cm	Less than 40 cm
Water pH at sunset	Below 9	Above 9
Dissolved oxygen before sunrise[1]	Above 3-4 mg/l	Below 3-4 mg/l

[1] Depending on the fish species present in the pond (see Section 25)

1. Because of the increased demand of dissolved oxygen caused by the addition of organic matter to the pond water, **you should limit the amount of animal manure to be applied at any one time**. This safe maximum amount is usually expressed in **kilograms (kg) of dry matter (DM) per hectare (ha) per day (d)** abbreviated as **kg DM/ha/d**.

2. **The safe maximum amount** of animal manure in cooler and temperate climates is 60 kg DM/ha/d or 0.6 kg DM/100 m^2/d; in warmer and tropical climates, the safe maximum is 120 kg DM/ha/d or 1.2 kg DM/100 m^2/d.

3. **To determine how much dry matter (DM)** a specific manure contains, you can **measure it yourself**. Take exactly 100 g of the manure you want to use and place it in a dry container such as a tin. Weigh the manure and the container (**W1** grams). Place the open container in an oven and heat it to about 180°C for four hours. Weigh the manure and container again (**W2** grams). The difference (**W1** − **W2** = **W3** grams) is **the moisture content** (in percent) of the manure. Obtain the **dry matter** weight **M** (in g) = 100 g − **W3**. This is also **the percentage of dry matter present** in your manure.

Example

You weigh 100 g of cow manure collected from a farm and place it in a tin. The total weight W1 = 185 g. After drying the sample in an oven, you find W2 = 145 g. The moisture content of the manure is W3 = 185 g − 145 g = 40 g. **The dry matter content** of your manure is DM = 100 g − 40 g = 60 g or **60 percent**.

4. Alternatively you can use **Table 17**, next to last column, which gives you the average dry matter content in percent of **the solid wastes**, for **fresh manures**. These values are only approximate and may be very different from the real value of the actual manure you want to use. For example, fresh cow manure usually contains 20 percent DM but dried cow manure may contain as much as 80 percent DM. It is thus always best to measure the DM content if you can.

25. Then, you can easily determine:

- **how much dry matter** a certain quantity of manure contains;
- the **maximum quantity** of a certain manure which you can safely place in your pond each day according to the above values.

Note: if you are using **liquid manure**, you should be even more careful because of its relatively high content in **ammonia**, a very toxic gas for fish. **Do not apply more than 1 000 litres/ha/d = 10 l/100 m^2/d.**

Examples

(a) 100 kg of fresh solid **pig manure** contains about 20 kg dry matter. If you have 375 kg, it contains 20 kg × 3.75 = 75 kg dry matter. In a cool climate, you can safely apply 60 kg DM/ha/d. Therefore in this case, you could daily apply the available pig manure to 75 kg DM ÷ 60 kg DM/ha/d = 1.25 ha of ponds.

(b) **Maximum amount of fresh solid manure** to be safely applied per day in 100 m^2 of tropical fish pond (see also **Table 17**):

Solid manure	Maximum amount (kg fresh/100 m^2/d)
Buffalo	6.3
Cattle	6.0
Horse	5.2
Sheep/goat	3.4
Pig	6.0
Duck	2.8
Chicken	4.8

Beware: if you do not use manure every day but only once a week, it does not mean that you can safely place on one day seven times as much manure into your pond. **The safe maximum amount remains the same**. If you wish to use more manure, you should reduce the interval between two consecutive applications. Place the manure two or three times a week.

26. The amount of animal manure to be applied to a particular pond **varies greatly, depending on factors** such as climate, water and soil quality, characteristics of the manure and kind of cultural system (type of fish, rearing density, length of rearing period). As for inorganic fertilizers, it is impossible to recommend any treatment valid under all circumstances.

27. For best results according to your particular case, you may apply one of the following procedures.

(a) As an approximate guide, **in small tropical rural ponds** generally from 100 m^2 to 300 m^2 in size, apply once or preferably twice a week, **one of the following**:

- poultry droppings, 4.5 kg/100 m^2;
- sheep or goat dung, 3 kg/100 m^2;
- pig dung, 6 kg/100 m^2;
- cattle or horse dung, 5 kg/100 m^2;
- cattle or horse stable-litter, 15 kg/100 m^2.

(b) For more controlled application **in a warm climate, at relatively low rearing densities of fish**:

- about ten to 15 days before stocking the fish, apply fresh manure at the equivalent rate of **10 to 20 kg dry matter per 100 m^2**, either to the dry bottom (drainable pond) or to the water (undrainable pond);
- about one week after stocking the fish, apply fresh manure to the full pond at the equivalent rate of **0.5 to 1 kg dry matter per 100 m^2**;
- every week thereafter, check on **water quality** (see above), increasing or decreasing the preceding amount of dry matter accordingly.

(c) **In a warm climate, at higher rearing densities of fish**:

- about ten to 15 days before stocking the fish, apply fresh manure at the equivalent rate of **10 to 20 kg dry matter per 100 m^2**;
- a few days after stocking, apply fresh manure at the equivalent dry matter rate per 100 m^2 (**Table 16**) equal to about **one-tenth the weight of fish stocked per 100 m^2**;
- every day thereafter, or at least twice a week, apply manure to the pond according to water quality and fish behaviour (see paragraph 19), **increasing or decreasing the amount of dry matter** applied. In general, you should gradually increase the amount as the total weight of fish present in the pond increases, until you reach **the maximum amount of dry matter** which can be safely applied on any one day, or about 1.2 kg/100 m^2.

Example 1

In the tropics, you wish to manure a 400-m^2 pond stocked with a **low density of fish**, using fresh chicken manure containing 25 percent dry matter. You may proceed as follows.

(a) **Ten days before stocking**, apply fresh chicken droppings at the equivalent rate of **15 kg/DM/100 m^2** or 15 × (100 ÷ 25) = 60 kg droppings per 100 m^2. You will need 4 × 60 kg = 240 kg droppings to manure your pond.

(b) **Seven days after stocking**, you measure SD transparency = 60 cm. Apply fresh chicken manure to the pond at the equivalent rate of **0.5 kg DM/100 m^2** or 0.5 × (100 ÷ 25) = 2 kg droppings per 100 m^2. Apply 4 × 2 kg = 8 kg droppings to the whole pond.

(c) **Seven days later**, you measure an SD transparency = 55 cm. Manure the pond at the increased equivalent rate of **0.8 kg DM/100 m^2**.

(d) **Every week**, check on water quality and adjust manuring.

Example 2

the tropics, you wish to manure an old 250-m² pond stocked with kg of Nile tilapia fingerlings using fresh solid **pig manure** containing percent dry matter. You may proceed as follows.

Ten days before stocking, apply the pig manure at the equivalent rate of **10 kg DM/100 m²** or $10 \times (100 \div 20) = 50$ kg pig manure/ 100 m². To manure your pond, you will need 2.5×50 kg = 125 kg fresh pig manure.

Three days after stocking, you measure SD transparency = 50 cm. Manure the pond at the equivalent rate of dry matter/100 m² equal to 1:10 of the fish weight/100 m². As there are 10 kg fish/ 250 m² = 4 kg fish/100 m², you need $(4 \text{ kg} \div 10) = $ **0.4 kg DM/100 m²**. You will need $0.4 \times (100 \div 20) = 2$ kg pig manure per 100 m² or $(2 \text{ kg} \times 2.5) = 5$ kg pig manure for your pond.

If you have enough manure, apply 5 kg pig manure **each day** thereafter for two weeks, carefully checking the water quality and fish behaviour in the pond.

After two weeks, if the water quality permits, slightly increase the daily amount of pig manure, for example up to 6 to 6.5 kg/day. If necessary, adjust this amount during the next days.

Every two weeks, slightly increase the daily amount of the manure until you reach **the safe maximum** of 6 kg/100 m²/day or $(6 \text{ kg} \times 2.5) = 15$ kg/day for your pond.

Mixing animal manures

28. If you do not have enough of one kind of animal manure, you may have **to mix two kinds of manure** to obtain **the total amount of dry matter** required for your ponds.

29. Calculate **how much of the second manure (M_2 in kg) you require as:**

$$M_2 = [DM_{tot} - (M_1 \times DM_1)] \div DM_2$$

where M_1 = available quantity of first manure, in kg;
DM_1 = dry matter content of first manure, in percent;
DM_2 = dry matter content of second manure, in percent;
DM_{tot} = total dry matter needed to manure the ponds, in kg.

Example

The ponds of your fish farm cover an area of 2 ha or $2 \times 10\,000 \text{ m}^2 = 20\,000 \text{ m}^2$. To manure the pond bottoms properly, you require 10 kg DM/100 m² and **a total dry matter** of $10 \text{ kg} \times (20\,000 \text{ m}^2 \div 100 \text{ m}^2) = $ **2 000 kg**. This is equivalent to $2\,000 \times (100 \div 25) = 8\,000$ kg of chicken droppings, at 25 percent dry matter content. However, you can obtain only **3 000 kg of chicken droppings**, and the rest of the manure has to be solid **pig waste**, 20 percent dry matter content.

Calculate **how much pig waste (M_2)** you will need to mix with the chicken droppings before treating your ponds, as follows:

- M_1 = 3 000 kg chicken droppings;
- DM_1 = 0.25 (25 percent) for chicken droppings;
- DM_2 = 0.20 (20 percent) for pig manure;
- DM_{tot} = 2 000 kg;

and therefore $M_2 = [2\,000 \text{ kg} - (3\,000 \text{ kg} \times 0.25)] \div 0.20 = 1\,250$ kg $\div 0.20 = $ **6 250 kg**.

Applying animal manures to a drained pond bottom

30. The dry bottom of a drained pond should be manured **at least two weeks after it has been limed** (see Section 54). **Inorganic fertilizers** may be applied at the same time, except if they contain too much calcium such as basic slag (see Section 62).

31. It is usually best **not to spread** the manure all over the pond bottom but instead **to stack it in heaps or in rows** at regular intervals.

Note: a number of methods for the placement or the distribution animal manure in various situations are illustrated on this and the following pages (pages 188 to 193). However, these illustrated examples are general in nature and must be adapted according to local condition (quality and quantity of manure available, water quality, weather, etc

**If the water flow can be controlled,
stack the animal manure at regular intervals
over the pond bottom ...**

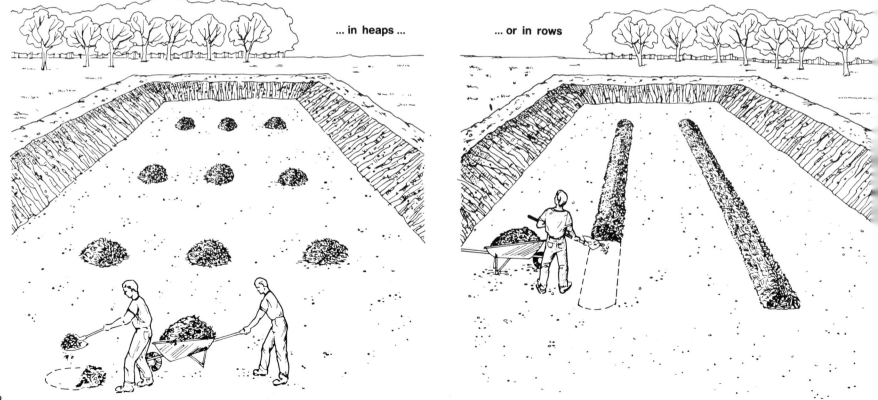

... in heaps ...

... or in rows

32. In new ponds with a sandy bottom, the manure should be spread all over the pond bottom area and worked into the surface soil. It is preferable to use either cattle and/or horse manure with stable litter or chicken droppings.

33. As soon as the manure has been applied, start filling the pond with water.

If the water flow can not be controlled, and the water exchange might be too great, stack the animal manure at regular intervals along the pond banks ...

... however in new ponds with a sandy bottom, spread the animal manure over the entire bottom area and work it into the surface soil

189

Applying animal manures to water-filled ponds
that have not yet been stocked

34. If the fish have not yet been stocked in the pond, pure animal manure may be applied evenly over the entire water surface.

35. However, if it is manure mixed with stable-litter, it is best to stack it in heaps along the banks. Mix these heaps from time to time.

**Applying pure
animal manure
from a boat**

Pond not yet stocked

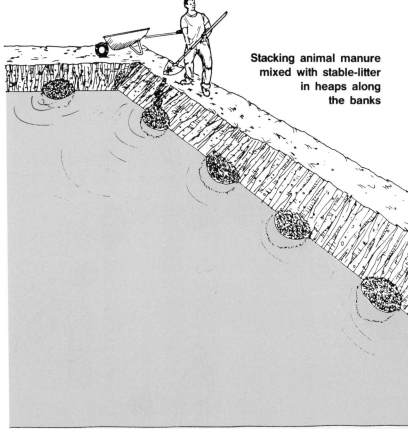

**Stacking animal manure
mixed with stable-litter
in heaps along
the banks**

Pond not yet stocked

36. Once the pond has been stocked with fish, you may apply **liquid manure** directly over its entire water surface.

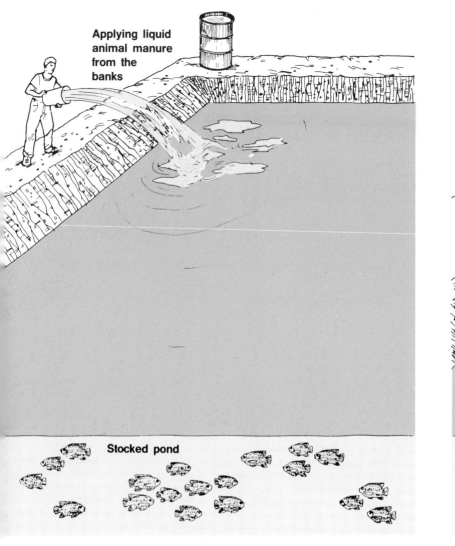

Applying liquid animal manure from the banks

Stocked pond

37. However, **solid manure** should only be applied indirectly. Use one of the following methods:

(a) A **floating basket** attached to the inside of a car tyre inner-tube similar to the one shown in item (b) on page 177. Prepare the amount of manure to be applied to the pond. Fill the basket with some of this manure, place it in the inner-tube and attach two long ropes. Pull the floating basket from bank to bank across the entire water surface. The manure will soften and gradually dissolve in the water. Refill the basket when necessary until all the manure has been distributed.

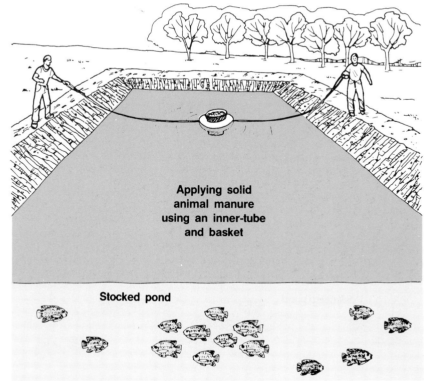

Applying solid animal manure using an inner-tube and basket

Stocked pond

(b) **Manure heaps**. Place the manure underwater in heaps at regular intervals along the pond banks.

(c) **A long crib** built next to the dike on one or more sides of the pond. you are using this kind of crib (see Sections 63 and 64), spread a mix the manure the full length of the crib.

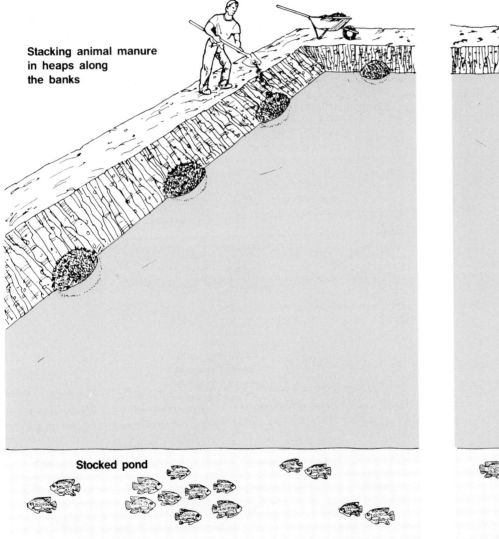

Stacking animal manure
in heaps along
the banks

Stocked pond

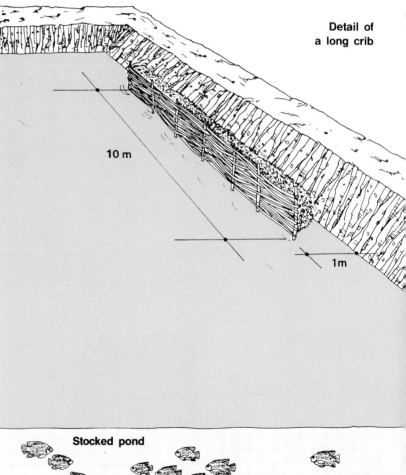

Detail of
a long crib

10 m

1m

Stocked pond

d) In smaller ponds, it is best to **build a crib** in each of the two shallow corners of the pond. Fertilize the cribs as you were told in item (c) on page 192.

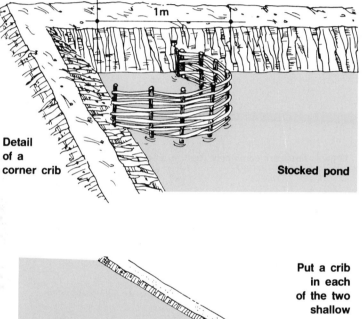

Detail of a corner crib

Stocked pond

Put a crib in each of the two shallow corners

38. **When planning the fertilization** of your ponds, remember the following:

(a) It is preferable to use animal manure as fresh as possible. Check if it is available when you need it.

(b) There should be at least a 15-day interval **between liming and manuring**.

(c) Apply manure preferably in **the early hours of the day**, about two to three hours after sunrise.

(d) Best results are obtained by **combining manuring with inorganic fertilization**. Additional phosphorus and nitrogen are usually beneficial to maintain a good plankton bloom (see Section 61).

(e) Always maintain and adjust your fertilization by checking water quality (see paragraph 19) and fish behaviour (see Section 25).

Note: remember that if you raise farm animals, it is an advantage to **keep them close to your fish ponds**. In some cases, you may even raise the animals in the pond, which is then automatically and continuously manured without extra cost or work. You will learn more about how **to integrate animal husbandry with fish farming** in Chapter 7.

63 Other organic fertilizers

1. Several organic fertilizers other than animal manures are commonly used for small-scale fish farming. These fertilizers are usually **wastes that can be obtained cheaply from local sources**. Part of the material is directly consumed by the fish, while the rest boosts the growth of the various **food-web** organisms used as food by the fish.

2. Among the most commonly used organic fertilizers other than animal manures, you can find:

- **slaughterhouse wastes** such as cattle rumen contents, blood, bones and enriched waste water;

Note: it is best to use **blood** for direct fish feeding (see Section 106).

- **agro-industrial wastes** such as cottonseed, molasses, mahua oilcake (see Section 47) and oil-palm sludge (4 to 5 percent nitrogen).

Note: wastes such as rice hulls, sugar-cane stalks and sawdust are rich in **cellulose**, which decomposes very slowly in the pond. Do not use them too much, unless you are trying to establish a good pond bottom on sandy soils.

- **biogas slurry**, the digested sludge remaining after biogas production, which contains about 10 percent dry matter;
- **cassava tubers** of the bitter species, which may be soaked in ponds to remove the toxic cyanhydric acid before consumption, form an excellent and cheap way to fertilize your small ponds;
- **aquatic vegetation** cropped from the pond itself or from the water canals or other water bodies; in some areas, floating plant pests such as the water hyacinth (*Eichornia crassipes*), water ferns (*Salvinia* sp.) and water cabbages (*Pistia* sp.) may be profitably used on your fish farm (see opposite page and also Section 64 on how to use them to make compost);
- **compost** produced outside the ponds, which (see Section 64) can be spread on the dried pond bottom before refilling or can be used to fertilize the water regularly.

3. **The average amounts** of these organic fertilizers to be applied t small ponds are given in **Table 18**. Apply them regularly, avoidin overloading the pond with several weeks' supply. Check the wate quality to control the quantities used. With the exception of slaughte house wastes and cassava tubers, these organic fertilizers are applied t the pond water in one or more **heaps**. You can also use either long c corner cribs as you were shown on pages 192 and 193. The fertilizin material is stacked and compacted inside to initiate underwater com posting (see Section 64). As described above, it is best **to turn over or least mix** the heap of decomposing material every week, before addir new organic matter.

TABLE 18

Organic fertilizers commonly used in small-scale fish farming

Organic fertilizer	Average amount applied at regular intervals
Animal manures	See Section 62
Cattle rumen content (slaughterhouse wastes)	10 kg/100 m²/week
Cottonseeds for in-pond composting (Agro-industrial wastes)	8 kg/100 m²/week
Biogas slurry	10-15 kg/100 m²/week
Cassava tubers fermentation	50-100 kg/100 m²/week 10-25 kg/100 m²/day
Vegetation for in-pond composting (soft, aquatic or terrestrial)	20-25 kg/100 m²/week
Compost (see Section 64)	20-25 kg/100 m²/week 50 kg/100 m² pond bottom

4. **Vegetation** such as cut grasses, vegetable wastes and rotting fruits may be used for simple **composting** (see Section 64) in the pond itself.

(a) Build a frame of wood or bamboo about 2 m × 2 m depending on the size of the pond.

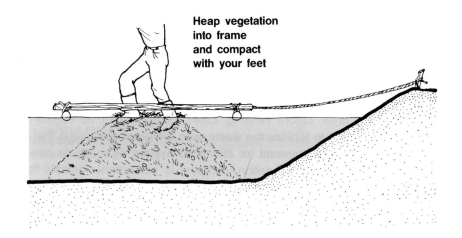

Heap vegetation into frame and compact with your feet

Wood or bamboo frame

(b) Attach it in the shallow end of the pond, about 1 m from the pond bank.

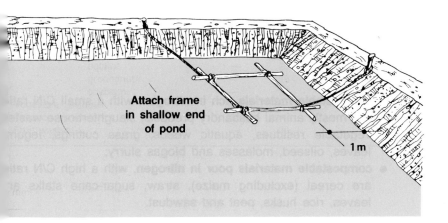

Attach frame in shallow end of pond

1 m

(c) Chop 200 to 250 kg of vegetal matter into small pieces, per 100 m² of pond water.

(d) Heap all this vegetation into the frame so that it decomposes underwater. Compact the heap well by trampling it with your feet.

(e) After seven to ten days, turn the heap over and remove all undecomposed pieces that have remained hard.

(f) Add finely chopped vegetal matter on top of the heap and compact it well. You will need 20 to 25 kg/100 m² pond.

(g) Seven to ten days later, turn the heap over, remove all hard pieces and add another 20 to 25 kg/100 m² of finely cut vegetal matter.

(h) Repeat this procedure about every week.

195

Making a good C/N mix

9. Some organic materials such as soybean stalks and leaves, groundnut shells, old grasses and weeds and some fruit wastes with a **good C/N value of 30 to 40** can be used directly for composting. In most other cases, several different types of material with **low and high C/N ratios need to be mixed** in the right proportions to give suitable C/N ratios.

10. To mix in the right proportions, proceed as follows.

(a) Make a list of the materials you plan to use for making compost. Obtain their C/N ratio from **Table 19**.

(b) Group them into two categories according to their C/N ratio, **a low C/N** less than 40 or a **high C/N** greater than 40.

(c) If you use more than one material in one of these groups, calculate **the average C/N** of that group for the quantities of each material available.

(d) Calculate **the proportions (on a dry weight basis)** in which you should use the materials from each group to have C/N within the 30 to 40 range.

(e) Calculate how much material of each group should be used to mix together according to their respective **dry matter content**, depending on their respective moisture.

11. Composting **materials with a very high C/N**, such as sawdust or wood chips, not only can be activated by adding materials with a low C/N but also by adding:

- **mineral nitrogen** (see Section 61); and
- **rock phosphate** (see Section 61);

at the rate of about **1 percent** of the weight of the high C/N raw materials.

12. **Do not add lime or ashes** to the composting material as this may increase the losses in nitrogen as the pH increases.

Example

You plan to use pig dung (C/N 14), chicken droppings (C/N 10) and rice straw (C/N 110) to make compost.

You have two groups of materials:

- low C/N: pig dung and chicken droppings; and
- high C/N: rice straw.

In the first group, you plan to use 30 percent pig dung and 70 percent chicken droppings. **The average C/N** of the group equals $(14 \times 0.30) + (10 \times 0.70) = $ **11.2**.

Calculate the proportions to be mixed from each group using **a simple graph** (see opposite page) based on:

- C/N group 1 = 11.2;
- C/N group 2 = 110;
- best C/N range = 30 to 40.

For best results, you should compost (**on a dry weight basis**) using a mix made of:

- pig dung and chicken droppings mixed in the proportion of 0.3 to 0.7;
- 70 to 80 percent of this animal manure mix and 20 to 30 percent of rice straw.

If **the average dry matter content** of the pig dung is 20 percent, that of the chicken droppings 25 percent, and that of the dried rice straw 80 percent, then you will need the following materials **for each 100 kg of composting material**:

- **animal manure mix** (75 percent) = 75 kg of which, 75 kg × 0.3 = 22.5 kg dry weight **pig dung**, equivalent to 22.5 kg × (100 ÷ 20) = 112.5 kg fresh pig dung; and 75 kg × 0.7 = 52.5 kg dry weight **chicken droppings**, equivalent to 52.5 kg × (100 ÷ 25) = 210 kg fresh droppings;
- **rice straw** (25 percent) = 25 kg × (100 ÷ 80) = 31.25 kg fresh rice straw.

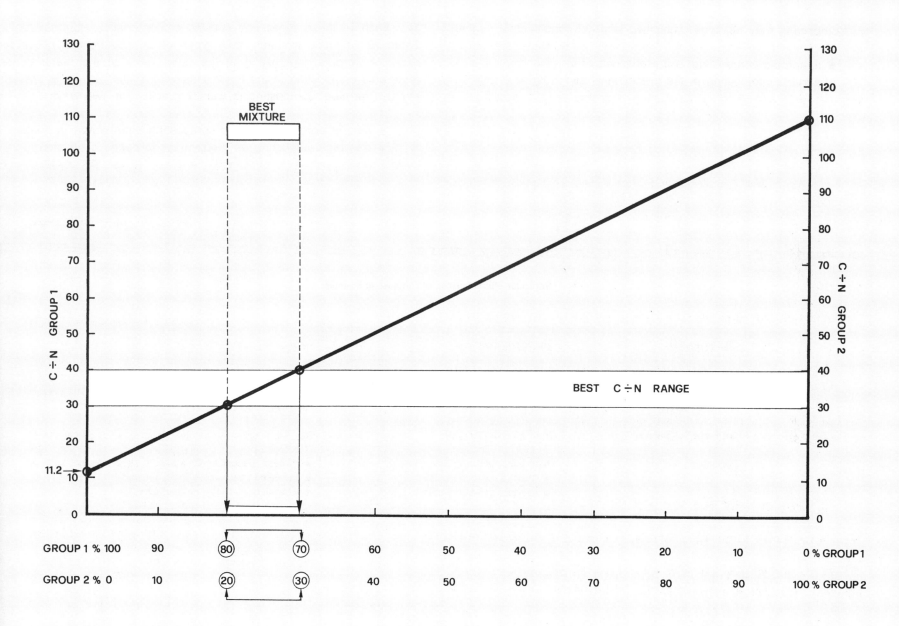

13. Composting may be done under two kinds of conditions:

- **anaerobic composting** in the absence of air;
- **aerobic composting** in the presence of air.

14. Each of these types has specific characteristics, summarized in **Table 20**. In some systems, both types of composting can occur, for example aerobic composting in the outer parts of the material and anaerobic composting in the inner area where little oxygen is present. For fish farming, composting is usually practised in two ways:

- simple aerobic/anaerobic composting underwater, in heaps;
- aerobic composting on land, either in heaps or in pits.

TABLE 20

Particular characteristics of composting methods

Characteristics	Aerobic composting	Anaerobic composting
Presence of oxygen	Necessary, for oxydation process	No, for reduction process
Losses of nitrogen	Yes, important (NH_3)	Reduced, if good sealing
Losses of carbon	Yes, important (CO_2)	Reduced, if good sealing
Production of heat	Important (up to 65°C or more)	Very small
Destruction of pathogens	Yes	No
Moisture content	To be controlled; best 40-60%	Not important
Composting method	• in heap, above ground level • in pit, below ground level • in heap, at water surface	• in heap, deeper under water • in sealed heap, above ground level • in sealed pit, below ground level

Preparing compost in the pond (aerobic/anaerobic)

15. You have already learned how to compost **aquatic and terrestrial vegetation** in the pond itself, where it can be heaped into a bamboo frame. Long cribs or **corner cribs** in the pond are also commonly used for such purposes.

16. You may use a simple material, such as grass or cottonseed; if one of the materials such as rice straw has a high C/N ratio, it would be better to alternate layers of this material with layers of a low C/N such as animal manure, cottonseed and waste fruit or vegetables.

17. The heap should be **well compacted underwater**, for example by trampling each layer well. Build the heap somewhat higher than the water surface, as its height will slowly decrease. **Every week, add some new layers of material** to build it up again.

18. For best results:

- use at least **one compost heap** per 100 m^2 of pond;
- make the **total surface area of compost enclosures** equal to 10 percent of the pond area;
- **turn over** the heaps every two to three days;
- place them in **deep enough water**.

Bamboo frame

Long crib

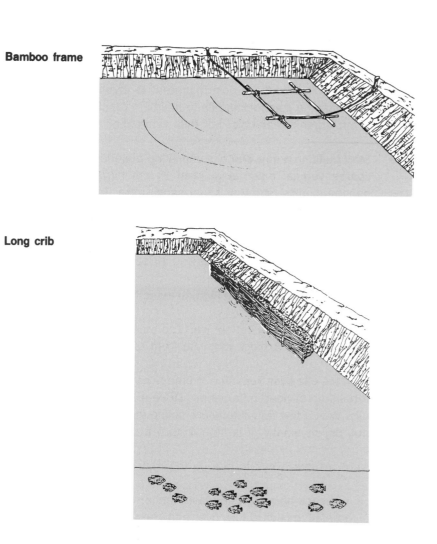

Corner crib

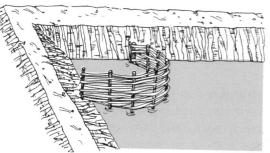

Preparing compost on land (aerobic)

19. If you decide to prepare compost on land, it is easier to use the aerobic method. It is then most important **to ensure that air is always present within the composting pile** to maintain rapid and full decomposition of the organic materials. For this, remember the following.

(a) **Start building a new pile** of composting materials with **a first layer of coarse vegetal material** at least 25 cm high, for example using banana stumps, straw or sugar-cane stalks. This layer should help air circulation while also absorbing any nutrient-rich liquid draining from the layers above.

(b) **Cut the composting materials** in small pieces of 3 to 7 cm.

(c) **Pack all materials loosely**, leaving air spaces between them. **Never compact this compost pile**.

(d) **Do not build your pile too high**, to avoid it compacting under its own weight.

(e) **Keep the pile moist** but not wet. Too much water results in a lack of air circulation. Protect your pile from rain (too wet) and from sun (too dry).

(f) **Turn the pile over** from time to time to aerate it and to avoid too high a build-up of heat in its centre. Drive a stick into the middle of the pile. Wait a few minutes before taking the stick out. **If the pile is too hot, dry or smelly**, it is time to turn it cver.

20. There are two ways to pile up the material:

- **in heaps above ground level**: better during seasons with heavy rainfall; easier to turn over and keep well aerated; but losses in nitrogen and carbon are higher; or
- **in pits dug into the ground**: site on high-level ground to avoid flooding; protect with trenches around if necessary; better in dry climates to retain moisture; losses in nitrogen and carbon are slightly lower.

21. **Remember** that you will **need water for composting**. To simplify the use of compost piles, it is better to **place them near a source of water**, for example a pond, and near **one of the main sources of composting materials** such as an animal shed.

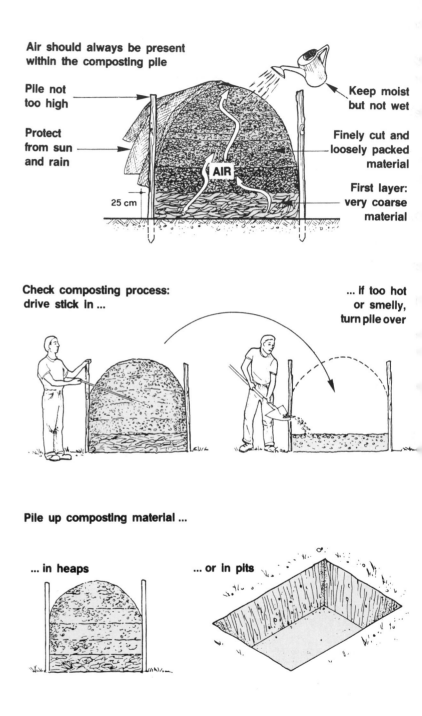

Air should always be present within the composting pile

Pile not too high

Protect from sun and rain

Keep moist but not wet

Finely cut and loosely packed material

First layer: very coarse material

AIR

25 cm

Check composting process: drive stick in ...

... If too hot or smelly, turn pile over

Pile up composting material ...

... in heaps

... or in pits

Preparing compost from grass cuttings

2. To prepare a cheap compost from **grass cuttings** under tropical conditions, you can use a simple **heap method** as follows.

a) Prepare your materials. You will need:

- 1 000 kg of fresh grass cuttings;
- 85 kg of good, dark soil; and
- 400 l of water.

b) Select a clean area and compact the surface soil. Mark the corners of a rectangular area 2 m × 4 m with four poles, 1.50 m above ground level.

c) Within this area, and starting from its perimeter, build a 40-cm layer of grass cuttings, about 250 kg.

d) On top of this, spread evenly about one-quarter of the soil. Incorporate it into the grass layer by shaking.

e) Moisten the layer with 100 l of water, using a watering-can.

f) Repeat this process three times to build a four-layer heap about 1.50 m high.

g) Protect the heap under a small shed built, for example, with bamboo and grass. It is best to close the sides with movable straw screens.

h) Turn the heap over about 25 days later. Rebuild the heap from 30-cm layers, watering each layer with about 100 l of water. Repeat this last process about 16 days later.

Depending on temperature and materials, your compost should be ready seven to eight weeks after starting.

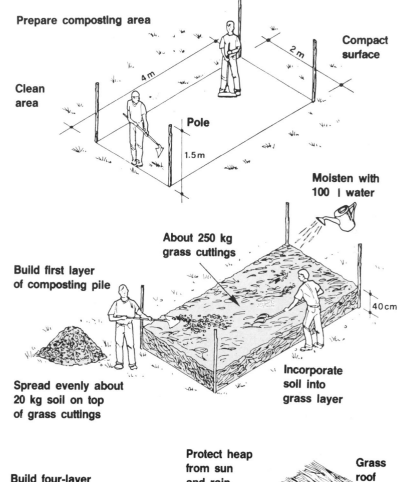

Prepare composting area

Clean area

Compact surface

Pole

2 m

4 m

1.5 m

Moisten with 100 l water

About 250 kg grass cuttings

Build first layer of composting pile

40 cm

Spread evenly about 20 kg soil on top of grass cuttings

Incorporate soil into grass layer

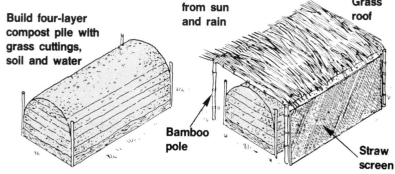

Build four-layer compost pile with grass cuttings, soil and water

Protect heap from sun and rain

Grass roof

Bamboo pole

Straw screen

Preparing compost from water hyacinth

23. Under tropical conditions, to prepare a cheap compost from **water hyacinth** (*Eichornia crassipes*), a floating aquatic plant, use **the heap method** as follows.

(a) Prepare your materials. You will need:

- 1 000 kg of fresh plants, sun-dried to give 400 kg of dried plants;
- 400 kg of fresh plants; and
- 12 bamboo pipes about 5 to 7 cm in diameter and 1.5 m long. To prepare the pipes, see Section 31, **Pond construction, 20/1**. Puncture the bamboo pipes along two opposite lines to give access to air;
- some coarse material, such as straw or banana stumps.

(b) Mix well together the dried and freshwater hyacinth plants.
(c) Mark the corners of a square area 3.5 m × 3.5 m with poles, 1.50 m above ground level.
(d) Within this area, build a first layer about 25 cm high with coarse material.
(e) On top of this first layer build a 40- to 50-cm layer with the plant mix.
(f) Insert the bamboo pipes vertically into this last layer, down to the first layer and about 0.8 to 1 m apart.
(g) Continue to build up the heap around the bamboo pipes with the rest of the plant mix.
(h) Protect the heap from rain and sun.
(i) After about 14 days, turn the heap over, watering it if it is too dry.
(j) About one month later, your compost should be ready.

Note: instead of bamboo you can also use stalks from plants such as corn tied in bundles of five or six stalks.

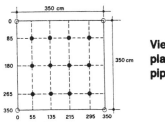

View of plan for placement of 12 bamboo pipes into pile

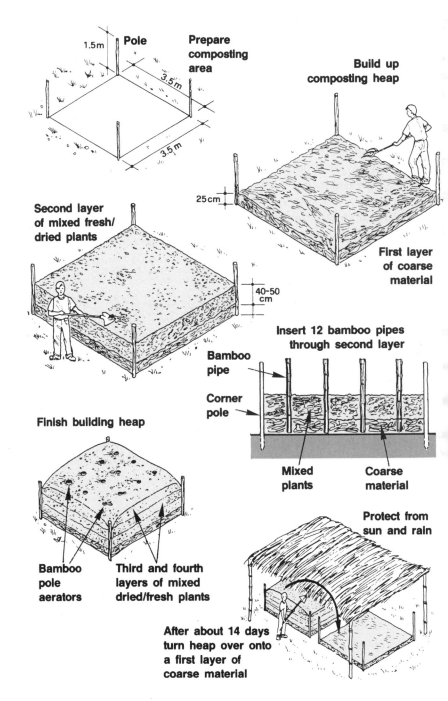

Prepare composting area

Build up composting heap

First layer of coarse material

Second layer of mixed fresh/dried plants

Insert 12 bamboo pipes through second layer

Bamboo pipe

Corner pole

Mixed plants   Coarse material

Finish building heap

Bamboo pole aerators

Third and fourth layers of mixed dried/fresh plants

Protect from sun and rain

After about 14 days turn heap over onto a first layer of coarse material

Preparing compost from plant materials and animal manure

4. A simple method of preparing compost in tropical rural areas **from various materials** such as tree leaves, grass, household wastes, rice husks, straw and animal manure, is the following:

a) Mark the corners of a square area 1.5 m × 1.5 m with poles about 1.5 m above ground level.

b) Within this area, build a first layer of coarse material about 25 cm high. Cut the rest of the material into small pieces.

c) Add a second layer of 5 to 10 cm made of **low C/N material**, preferably animal manure. Moisten as necessary.

d) Build up your heap until it is about 1.5 m high by adding more layers. Alternate 20-cm layers made of high C/N material with 10-cm layers made of low C/N material. If you do not have enough animal manure, sprinkle some **nitrogen fertilizer** on top of each high C/N layer (see Section 61). **Moisten each layer** so that it is damp but not soggy.

e) If it does not rain, you may need to sprinkle water on top of the heap every three days.

f) Turn over your pile after ten to 14 days. Check its heat production and moisture regularly.

g) Your compost should be ready after another ten to 15 days.

Note: you can protect the composting heap from sun and excessive rain by covering it with plenty of straw and by rounding off the top of the heap.

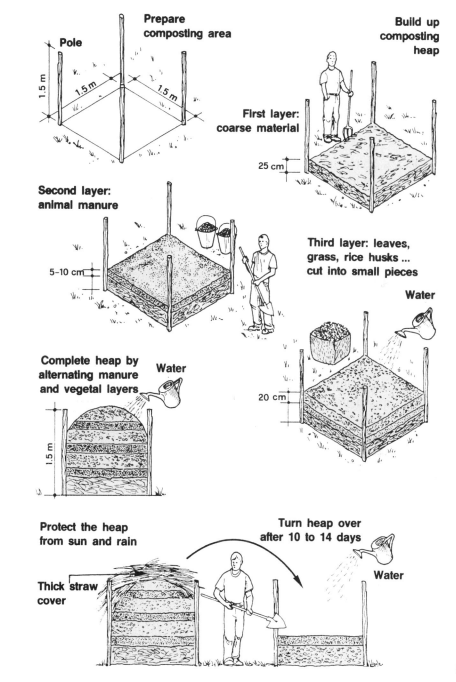

Prepare composting area

Pole

1.5 m

1.5 m

1.5 m

Build up composting heap

First layer: coarse material

25 cm

Second layer: animal manure

5–10 cm

Third layer: leaves, grass, rice husks ... cut into small pieces

Water

20 cm

Complete heap by alternating manure and vegetal layers

Water

1.5 m

Protect the heap from sun and rain

Thick straw cover

Turn heap over after 10 to 14 days

Water

Preparing compost using the pit method

25. With this method, **several pits are dug side by side**. They can be of any size, although it will be easier if they are not too deep. A practical size is 1.5 m × 3 m by 1.5 m deep.

26. To prepare compost so that it becomes continuously available, proceed as follows.

(a) At the bottom of the first pit, build a first layer of coarse material, 30 cm deep. Chopped banana stems, banana tree leaves and straw are good.

(b) Add a second layer about 10 cm thick of a mix of low C/N material, including animal manure if available, and moisten as necessary.

(c) Set up, or lay across, bundles of grass, stalks or sticks, or bamboo pipes (see above), to provide air spaces in the composting pile.

(d) Add more layers, alternating 20-cm layers of high C/N material with 10-cm layers of low C/N material, moistening each layer as necessary. Add some more air spaces across.

(e) When the pit is full and the pile rises above the ground, cover it with plenty of straw.

(f) Sprinkle water over the pile as necessary every three days.

(g) After about three weeks under tropical conditions, turn the pile over by transferring it to an empty pit next to it. Moisten as necessary and cover the new pile with straw.

(h) Start composting in the first empty pit while checking the process in the second area.

(i) Depending on the climate and the kind of materials used:

- either your compost will be ready a few weeks later;
- or after a few weeks you will have to turn the pile over again, by throwing it into a third pit.

(j) Transfer the pile from the first pit into the second pit.

(k) Start composting in the first pit again.

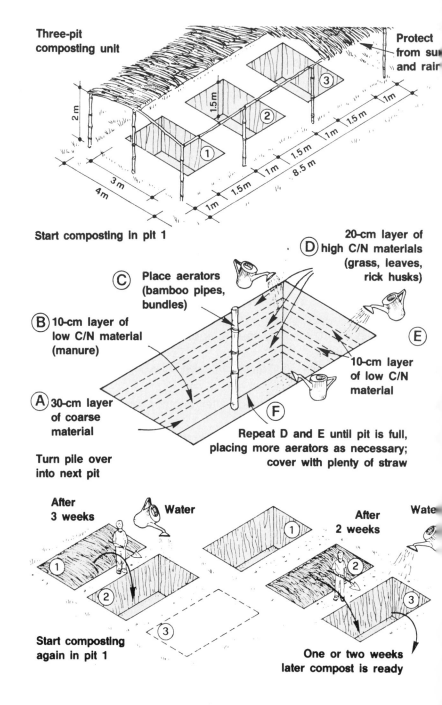

206

1. In the last chapter you have already learned a great deal about animal manures as organic fertilizers (see Section 62) and about their advantages for fish production. In this chapter you will learn more about the animals themselves and how best to integrate their production with fish farming.

2. If you are already raising some livestock on your farm, and have suitable conditions for ponds, you may be able to combine animal husbandry with fish production. This can help you to increase your income at little extra cost and, at the same time, provide a valuable solution to the problems of manure disposal. But if you farm fish and are **not yet familiar with animal husbandry**, you should be aware that starting such an activity, and integrating it with fish production, can be quite difficult and may need some time to work properly.

Additional inputs of animal husbandry

3. As a fish farmer, if you want to raise livestock, you will need a number of additional inputs to integrate animal husbandry and fish farming. Some of these are:

● more land and more water;
● greater investments for housing, food and young animals;
● new skills for managing your livestock;
● extra work to manage it properly;
● probably better and greater quantities of feed.

4. However, because of the fluctuations in quality and quantity of the manures available, pond management may become more difficult. Water quality will have to be closely and more frequently monitored, particularly if the animals are manuring the ponds directly.

Beginning successfully

5. At the beginning, do not be too ambitious. **Start in a small way**, with one or two ponds and a few animals.

6. Concentrate on the **fattening of young animals** bought from other farmers. Later, you may decide to breed your own animals.

7. If you need advice **ask more experienced people** who already have integrated animal production with fish farming or perhaps your extension agent can help you.

8. Take very good care of your animals. They are an important and costly investment.

9. Ensure that an adequate **supply of good water** is always available both for your animals and your fish.

10. Confine your animals close to the ponds, at least at night.

11. Buy **locally available construction materials and feeding ingredients** at reasonable prices. Make full use of available land to produce protein-rich **plant material** for feed, erosion control (see Section 41) and fencing (see Section 45).

12. Closely check the **water quality** of your ponds and in particular its dissolved oxygen content (see Section 25). Manage the water supply accordingly.

13. **Adapt the feeding** of your fish to the amount of manure distributed (see Chapter 10, **Management, 21/2**).

Animal husbandry and fish farming interrelationships

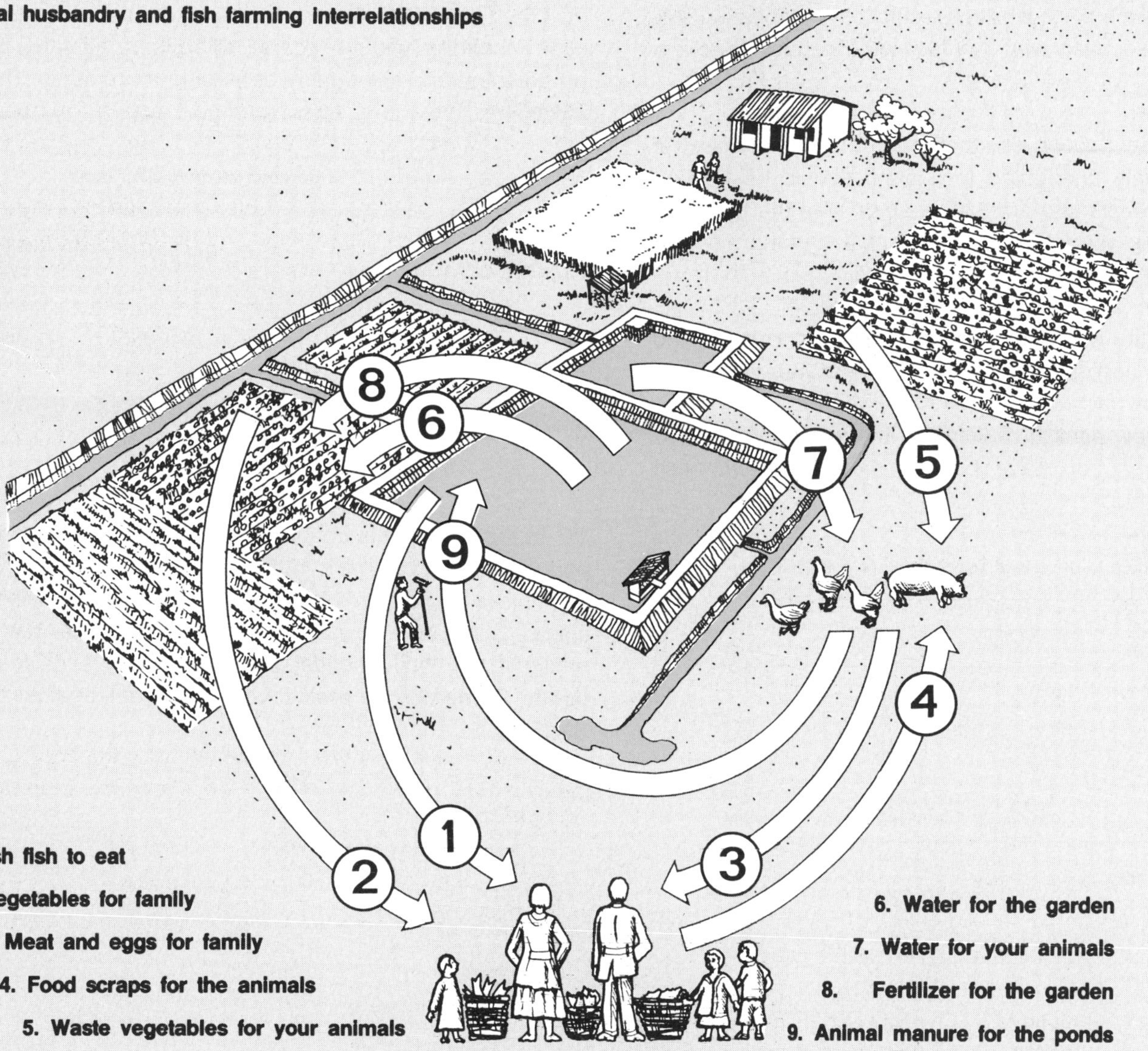

1. Fresh fish to eat
2. Vegetables for family
3. Meat and eggs for family
4. Food scraps for the animals
5. Waste vegetables for your animals
6. Water for the garden
7. Water for your animals
8. Fertilizer for the garden
9. Animal manure for the ponds

14. A number of animals, such as pigs, ducks, chickens, rabbits, sheep, goats and cows, can be associated with fish farming. The first three are the most popular.

15. Earlier you learned (see Section 62) that **the maximum amount of manure** which may be safely distributed to a pond should be limited, according to pond size, local climate and fish species. Use the following chart to estimate **how many animals** you will be able to integrate with farm fishing. However, remember that:

- you should select the average number of animals to be raised per 100 m^2 of pond according to the sensitivity of your fish to low oxygen content;
- initially, choose **a low density** of animals and increase densities carefully, only when you have enough experience with fish farming.

Number of animals to be raised per 100 m^2 of pond

	Low density	High density
Pigs	0.2-0.3	0.5-1
Ducks	3-5	15-20
Chickens	5-15	20-35
Sheep/goats,		
all day	2	4
night only	4	8
Oxen/cows,		
all day	0.1-0.2	0.3-0.4
night only	0.2-0.4	0.6-0.7

16. When selecting a particular animal to be integrated with fish farming, local conditions should receive priority, particularly concerning marketing potential and availability of feeding ingredients. **The duration of each production cycle** may also be important.

Duration of production cycle

	Weight	Production cycle
Meat duck	50 g to 1.5-3 kg	2-3 months
Broiler chicken	30-40 g to 1.2-1.7 kg	3-4 months
Pig	15-20 kg to 60-100 kg	5-8 months

17. In the rest of this chapter, you will learn more about the three most popular kinds of animals to be integrated with fish farming: pigs (Section 71), chickens (Section 72) and ducks (Section 73). Other animals, such as sheep, goats and cows, can also contribute to pond fertilization more easily if you build sheds, corrals, paddocks or feedlots **close to the ponds**. Remember to protect your dikes well against animal erosion (see Sections 44 and 45).

1 Integrated pig farming

A simple approach is to buy 15- to 20-kg piglets and to raise them for five to eight months, until they reach market size, which is usually when they weigh from 60 to 100 kg. Under tropical conditions, particularly in humid climates, every effort should be made to keep **the ambient air temperature** as cool as possible, especially when raising pigs larger than 70 kg. Select a breed well adapted to local conditions of climate and management.

Selection criteria of two European pig breeds

European pig breeds useful for the tropics are the **Large White** and **Duroc Jersey**. Thermal limitations are as follows:

(a) Average annual air temperature up to 25°C.
(b) Periods when air temperature is well above 30°C should be limited.
(c) Comfort zone and optimum ambient temperature:

- pigs 30 to 70 kg = 10 to 25°C (optimum 24°C);
- pigs 70 to 120 kg = 10 to 20°C (optimum 15°C).

Housing your pigs

It is important to house your pigs in an area which offers at least the following:

- proximity of an ample **water supply** for drinking and cleaning;
- good **ventilation** to reduce ambient air temperature whenever necessary;
- good **drainage** to avoid excess local humidity, especially at night;
- easy **access** for management purposes;
- good **manure collection** facilities.

3. Basically, pig manures may be distributed to fish ponds in one of two ways:

(a) **Direct transfer**, either immediately or through cleaning.
(b) **Indirect transfer** from a storage pit built next to the pig sty. Calculate the required **storage volume** according to the planned duration of storage as:

- for 20 to 45 kg live weight pigs: 0.004 to 0.006 m^3/day/pig;
- for 45 to 90 kg live weight pigs: 0.008 m^3/day/pig.

4. Although the **indirect method** results in a better-regulated distribution of the manure to a series of fish ponds, the **direct method** is often preferred by small farmers to avoid having to store and handle the manure. In this case, the pig sty can be built in one of three ways:

(a) **Completely or partly over the pond water**: The pig sty is usually built cheaply of local materials with a latticed floor to allow manure and uneaten feed to fall directly into the pond. However, the latticed floor may cause leg injuries, and high humidity from living above water may cause respiratory diseases. Pig and fish production cycles need to be matched well.
(b) **Partly on one of the pond dikes**: Built about 0.20 m above maximum water level, the pig sty gives access to a bathing area within the pond and the pigs will often defecate in the bathing area thus fertilizing the pond. The rest of the manure and uneaten feed are regularly washed into the pond while cleaning the pig sty. If waste materials are not needed for the pond, they must be moved elsewhere, possibly to another pond.
(c) **On top of one of the pond dikes**: The pig sty floor slopes either to a manure-collecting channel or directly to the pond. A **manure alley** is included in the pig sty to encourage the animals to defecate next to the collector or pond. Regular washing directs the manure into one or several ponds. With this system you can easily direct the available manure through a series of channels to the particular pond or ponds that require fertilizing.

Designing a pig sty for tropical conditions

5. A well-designed pig sty should remain relatively cool even in warm weather.

(a) **It should provide maximum shade** for the animals. Allow good ventilation space inside, but keep the roof edges low. Increase roof overhangs. A two-slope roof is preferable. Orient the pig sty if possible in an east-west direction and plant shading vegetation which does not affect ventilation.

(b) **It should be well ventilated**, particularly **at ground level**. Keep the construction open and avoid using solid walls. Orient it as much as possible across the direction of prevailing winds.

Note: if there are cold spells, remember that your pigs may then need to be protected from cold draughts by closing some of the existing openings. Simple flap or clip-on shutters can be useful.

6. Design your pig sty to make **its management easier**, in particular allowing for a regular distribution of feed and for cleaning with plenty of water.

7. Additional hints for good pig sty design are given here.

(a) In warm climates provide more **floor space**:

- 1 m² per pig of 20 to 50 kg live weight;
- 2 m² per pig of 50 to 100 kg live weight.

(b) The normal practical **size of pens** is 15 to 20 m² for most pond-side units. The area should be roughly square.

(c) **Do not use more than six** individual pens in a single row. If you buil more, use a **double-line design** with a 1.5-m central alley. Thi layout is much more expensive, however, especially for roofing.

(d) **A simple way to separate pens** from each other is to use 20-cr boards nailed to strong vertical posts. You may also use horizonta wooden poles or a 15-cm-thick wall made of bricks or concret blocks. The total height should be 0.9 m.

(e) For **easy access to each pen**, build strong, 60-cm-wide doors wi 20-mm-thick wood. Use strong hinges and a dropping lock.

(f) **Feeding troughs** should be fixed in place, easily accessible fr outside the pen, easy to clean and designed so as to limit fe losses. A bottom draining hole may be added. Cement is the be material, but simple wooden trays can also be used. The troug should be long enough to allow the pigs to feed without crowdi (see the chart below).

Suggested trough lengths for feeding pigs

Live weight of pigs (kg)	Minimum trough length (cm/pig)
20-30	18-20
30-50	22-24
50-70	25-27
70-100	30-32
about 100	35-37

(g) You will need from 2 to 7 litres of **drinking-water** per pig per da depending on its size. You can provide water in three ways:

- in the feeding trough, between meals;
- in a small channel of running water;
- in a drinking container.

h) **The floor of the pen** should be designed so that it can be easily scraped with a shovel and washed with a brush and plenty of water. It should be very strong. For easy drainage, build the floor with a 2 to 3 percent slope toward the front of the pig sty, where the manure will be collected. Preferably build it with:

- hard stones and strong mortar (see Section 33, **Pond construction, 20/1**);
- 10-cm-thick concrete (see Section 34, **Pond construction, 20/1**) placed on a stone or gravel foundation (at least 15 cm thick) and covered with a strong mortar and a cement screed. Make the surface rougher by marking a zig-zag pattern in it before it dries.
- **Remember** to adapt the number of pigs to the size of the ponds you want to fertilize with manure so that you do not overload your ponds with organic manure (see Section 70).

Some simple designs for a pig sty

Here are some simple designs for a pig sty you can build quite easily using mainly local materials.

) **To build a pig sty on the pond dike with easy access to a bathing area** in the pond itself, you will need: two pens of 2.20 m × 3 m = 6.60 m^2 each for four to five small pigs each and one pen of 2.20 m × 3.50 m = 7.70 m^2 for seven to eight small pigs. Total floor area is 20.90 m^2. The total capacity is 15 to 18 small pigs or eight large pigs. Provide drinking-water through running-water channels. **Bathing areas** (30 cm deep) are not covered and are paved with stone and mortar. Provide access with a sloped concrete ramp.

(b) **For a simple pig sty built on higher ground** with a brick collecting channel feeding the manure to the pond, sited on a lower level, you will need two pens of 3 m × 3 m = 9 m^2 each. The total floor area is 18 m^2. A metal roof is supported by wooden poles and six brick pillars, and there is metal mesh all around.

(c) **A better pig sty** can be built on a pond dike with a manure alley and drain toward either a manure storage pit or to one or several ponds. Build a two-sided thatched roof, shorter in front and above the manure alley. Use wooden poles and boards. Build a concrete pen floor, making each pen 3 m × 2.70 m = 8.10 m^2 plus 1.40 m × 2.70 m = 3.78 m^2 for a manure alley. Drinking troughs and feeding troughs should be constructed in cement.

Feeding your pigs well

9. Pigs have the advantage of being able to accept and make good use of a **wide variety of food** such as:

- kitchen wastes: household, restaurant, canteen;
- aquatic plants: water lettuce, water hyacinth;
- crop wastes: banana tree stumps, waste vegetable or fruit;
- slaughterhouse wastes;
- agricultural by-products: rice bran, oilseed cakes.

10. You will learn more about these various feeds later (see Chapter 10, **Management**, 21/2). Make sure to provide your animals daily with **plenty of vegetal food** in addition to any other types of feed. **Dry feedstuffs** should be mixed with water.

11. If you are growing **an improved breed**, such as the Large White, better production will be achieved by using better quality, **balanced feeds**, made from a mixture of several ingredients as shown below. This feed mixture sometimes includes a special concentrate of vitamins and minerals which is obtained from specialized suppliers. Distribute **the following rations** daily, in two to three meals, mixed with water, according to the live weight (in kg) of the growing pigs.

Daily food rations for growing pigs

Live weight	Daily ration
20-40 kg	1.50 kg/day
40-60 kg	2.00 kg/day
60-80 kg	2.50-2.70 kg/day
80-100 kg	3-3.20 kg/day

Three possible types of composite feeds for pigs (in percent fresh weight)

Type 1		Type 2		Type 3	
Wheat middlings	30	Wheat bran	33	Maize meal	44
Cassava	39	Cassava	33	Groundnut cake	15
Cottonseed cake	13	Blood	15	Rice polishings	30
Groundnut cake	13	Vegetal matter	19	Wheat meal	5
Fish meal	2			Concentrated	
Bone (calcinated)	2.5			vitamins	
Minerals	0.5			or minerals	6
Total	100.0		100		100

12. Distribute the food freely to make sure that there is always food available so that it can be consumed whenever the pigs want it.

13. The **food conversion ratio** should vary around 8:1 (see Section 103 **Management**, 21/2).

Remember: at least once a day, clean the pens and the feeding trays well, using plenty of water.

Integrated chicken rearing

A simple and commonly used method for rearing chickens is to obtain **one-day-old chicks** from a specialized supplier and rear them to:

- 10- to 12-week-old **broilers** of 1 to 2 kg each; or
- 4.5- to 5.5-month-old **laying hens**, which will be reared for 12 to 18 months to produce eggs.

In both cases, the first five-week rearing of the one-day-old chicks is similar (as described in paragraphs 6 to 10 in this section).

When rearing chickens with fish, remember the following points:

If you choose to produce batches of broilers, the amount of manure available from each batch will be very low at the beginning of the rearing cycle. It will increase during the growing period until reaching a maximum toward the end. Be careful then not to put too much manure into your ponds. Good planning will also be essential for the replacement of the chicken stock at short regular intervals, every five to seven weeks.

If you choose to raise laying hens for egg production, the supply of manure will be much more constant for a much longer period, between 1 and 1.5 years. The chicken stock will have to be replaced occasionally.

Quantity of manure produced according to live weight of chickens

- **a five-week-old chicken** produces about 1 g dry matter/day;
- **a 12-week-old broiler** produces about 45 g dry matter day;
- **a laying hen** (0.5- to 1.5-year-old) produces about 40 to 50 g dry matter/day.

Choosing the chicken breed

4. Select the chicken breed among those available locally according to the kind of production you plan.

(a) For the **production of broilers** you will probably have a choice between local breeds, improved breeds (for example, Sussex or New Hampshire) or their hybrids.

(b) For the **production of eggs**, improved breeds (for example, Rhode Island or White Leghorn) or their hybrids are usually preferred, although the above breeds can also be used for mixed productions of meat and eggs.

5. If possible, consult local specialists for the final choice.

Nursing one-day-old chicks

6. Select a well-drained site on which to build a **small brooding house**. It should have the following characteristics:

- floor area of at least 2.5 m^2 per 100 chicks to be reared;
- well ventilated but with no air draughts at ground level;
- protected from strong winds;
- shaded, subdued natural light;
- well protected against predators;
- easily cleaned and disinfected;
- close enough to your home for good care.

7. Proceed as follows.

(a) **Two weeks before** each batch of chicks arrives, clean, repair and disinfect the brooding house. Prepare the following for each brooding house:

 • **one feeding tray** (length 1 m) per 50 chicks;
 • **one drinker** per 50 chicks such as a 5- to 10-l can or bucket (sunk into the floor so that only about 10 cm is above floor level) or a shallow bowl supplied with water from an upside-down bottle;
 • **additional heating** to keep the chicks warm without crowding together. You may use two **kerosene storm lanterns protected by netting**. If necessary, an oil drum cut in half lengthwise can be suspended over the lantern(s) to direct heat down;
 • about 6 m of chicken wire 60 cm high to make **a circular pen** of about 1.80 m in diameter, enough to rear 100 chicks. Attach paper sheets or jute sacks to this low fence to protect the chicks against draughts.

(b) **Ten days before the chicks arrive**, prepare the necessary supplies:

 • **to feed** the chicks (see paragraphs 8 to 10 in this section), you will need about 900 to 1 000 g of special chick feed per chick (first age); if not included in the feed, you should also purchase some vitamin/mineral supplements;
 • **to protect** your chicks from diseases or to cure them if they become sick, ask a local specialist for more details, and obtain drugs and vaccines from a supplier.

(c) **One day before arrival**, add a good dry litter, such as wood shavings, in the brooding area. Make sure the kerosene lamps are functioning properly.

(d) **On arrival day and for the first week**:

 • place suitable **feed** in the trays;
 • place **water** in the drinkers;
 • bring the **air temperature** of the brooding area up to 34°C, lighting the kerosene lamps and protecting them well with fir netting;
 • check on **the fence** and its cover against air draughts;
 • gently place the one-day-old chicks within the brooding area
 • **during the following hours** and especially at night, check oft the behaviour of the chicks and maintain air temperature clo to 34°C at all times, using the half-drum brooder as necessar

Note: you should never mix chicks of different age groups.

(e) **Second week**:

 • reduce air temperature to 32°C in the brooder;
 • remove the barrier encircling the brooding area;
 • add some feeding trays and drinkers, if necessary, and add o tray with very small gravel and one tray with **limestone* g** (calcium).

(f) **Third week**: reduce air temperature to 30°C.

(g) **Fourth and fifth weeks**: reduce air temperature to 28°C to allow t chicks to slowly adapt to local temperature variations.

(h) **Sixth week**: transfer the young chickens to an outside reari house. Do not use the brooding area for another 15 days.

Note: in addition, make sure that there is **always food and wat** available in the brooding area; also, change the litter and clean a disinfect the brooding area once a week.

Give your chicks as much dry feed mixture as they can eat.

You can buy the mixture as **chick feed** or **first age** chicken feed from specialized suppliers. It should contain all the necessary ingredients, including vitamins and minerals, for successful chick rearing.

0. You can **mix several ground ingredients** yourself (see the following chart). Vitamins and minerals are either included in this mixture or have to be given separately.

**Three types of mixed feeds (in percent of fresh weight)
to be used as chick feed**

Ingredients	Types of feed		
	1	2	3
Maize or millet	60	56	60
Groudnut cake	25	29	—
Wheat middlings	—	—	25
Blood meal	10	—	—
Bones, calcinated/ground	4	—	—
Salt	1	—	—
Concentrated mix (proteins/vitamins/minerals)	—	15	15
Total	100	100	100

Producing broilers and fish together

11. Chickens grow better at moderately warm temperatures, as their comfort zone ranges from 13 to 24°C. At temperatures above 30°C, production may be reduced, especially when humidity is high. In tropical climates, their environment will therefore have to be kept as cool as possible.

12. **Select a site** with good ventilation, but protected from strong winds. **Design your chicken house** with the following features in mind.

(a) Orient it to the extent possible east-west, against direct sunshine.
(b) Orient it to the extent possible at right angles to **dominant winds**. Limit its width to permit good ventilation across the building.
(c) Use **low walls** all around to protect at ground level.
(d) Use wire mesh above (18- to 25-mm mesh).
(e) Keep **predators**, rodents and birds out.
(f) Use corrugated steel sheets or thatch for the roof, with at least 50 cm overhang all around.
(g) Build a tight-fitting door.
(h) Fix reed curtains on the windward side, if necessary.
(i) Allow at least **0.125 m² floor space** per growing chicken.

13. **Build your chicken house** close to the fish pond in one of two ways.

(a) **Over a fish pond**: with access via a removable gangway, the house should have a wooden slatted or wire mesh floor, allowing wastes to drop directly into the pond. Limit the total number of animals according to pond size, climate and fish species (see Section 70). This design is preferred for rearing laying hens, which supply the pond with a relatively constant quantity of manure (see above).
(b) **On one of the pond dikes**: the house should have a well-drained floor made of compacted soil, gravel, wood or concrete. Some dry litter may be used, according to the needs of the pond, to collect the droppings. This system is better where variable amounts of waste are produced.

14. Provide the following equipment in the chicken house:

- **feeding trays**: one or two 1-m-long trays per 20 animals;
- **drinkers**: one or two 5- to 10-l bowls per 50 animals;
- **trays with small gravel and limestone grit**;
- **perches**: preferably placed lengthwise along the centre of the house should be about 3.5 cm in diameter and fastened to solid stands about 60 cm above the ground. Provide 10 to 15 cm of perch per animal. It is best to attach a **manure-collecting deck** about 20 cm underneath, if the house is not built over a pond.

15. **To produce broilers successfully**, proceed as follows.

(a) When they are five weeks old, transfer a batch of young birds from the brooder house to the chicken house. **Never mix together animals of different ages or breeds**. Allow a density of eight birds/m^2 or less.

(b) Ensure that there is always good **drinking-water** available.

(c) **Feed the broilers well**, as follows:

- **sixth week**: use the same food as for chicks (see paragraphs 8 to 10 in this section), at the rate of 50 g/day/bird;
- **from the seventh week on**: use a broiler feed (see following chart) at an increasing rate of 70 to 140 g/d/bird;
- provide **fresh vegetal matter** every day;
- give **a vitamin/mineral supplement** (for example, 2 ml/d/10 birds in the drinking-water), if necessary.

(d) Administer **drugs and vaccines** as recommended by local specialists.

Three possible types of broiler feed mixes
(in percent of total fresh weight)

Ingredients	Types of feed		
	1	2	3
Maize or millet meal	71	60	60
Rice polishings	—	20	—
Brans (wheat, rice)	—	—	20
Groundnut cake, milled	20	7	—
Coprah cake, milled	—	—	7
Blood meal	5	—	—
Bones, calcinated/ground	3	—	—
Salt	1	—	—
Concentrated mix (proteins/vitamins/minerals)	—	13	13
Total	100	100	100

16. Under favourable conditions of temperature, humidity and feeding the food conversion rate should remain under 3:1 (see Section 10: **Management, 21/2**).

17. Broilers are usually culled when they reach about 12 weeks of age. They should then each weigh from 1 kg (subtropical climate) to 1.7 kg (tropical climate). When the chicken house is empty, clean and disinfect it well. Leave it empty for at least 15 days, if possible, to help control diseases.

18. **Select the site** and **design your chicken house** as described in paragraph 13 for producing 12-week-old broilers, noting that **more floor space** will be required for laying hens, at least 0.33 m² per bird. Limit the width to 2 m.

19. You may build the chicken house either:

- **over a fish pond**: using a wooden slatted or wire mesh floor (limit the total number of animals according to pond size, climate and fish species, see Section 70); or
- **on one of the pond dikes**: preferably using a wooden or concrete floor and a deep dry litter to collect the droppings.

20. In the chicken house, provide the following equipment:

- **feeding trays**: two 1-m-long trays per 20 hens, using the same model as for broilers until the age of about five months and then replacing it with a **standing trough**, providing at least 10 cm of trough length per bird;
- **drinkers**: two 10- to 15-l buckets fixed on a stand, per 20 hens;
- **perches**: same as for broilers until the age of about five months and then replacing them with larger perches (preferably 2.5 × 4 cm section or 5-cm diameter bamboo), providing at least 25 cm of perch per bird (if the house is not built over a pond attach **a manure-collecting deck** about 20 cm below);
- **nest boxes to lay eggs**: along one of the house walls when hens are ready to start laying (14- to 23-week-old), providing one single nest for every five birds, with 10 cm of **dry litter** (maize leaves, grass, rice stalks) in each nest;
- **gravel tray**: one 80-cm tray containing 4 to 6 mm gravel to help the hens digest their food better;
- a similar sized tray with **limestone grit** (calcium).

21. **To rear laying hens and produce eggs** successfully, proceed as follows.

(a) When they are five weeks old, transfer a batch of young birds to the chicken house. **Never mix together** animals of different ages or breeds. **Maximum density** is eight birds/m² of floor space.

(b) Ensure that there is always good **drinking-water** available.

(c) **Feed them well** early in the morning and, if necessary, early afternoon:

- from **the sixth to ninth week**: chick feed (see paragraph 15 in this section) at the rate of 55 g/day/bird;
- from **the tenth week to first laying**: broiler feed at an increasing rate of 70 to 95 g/day/bird;
- **egg-laying period**: layer feed (see following example) at an increasing rate of 105 to 125 g/day/bird during the first four weeks and then at the constant rate of 125 to 130 g/day/bird;
- provide **fresh vegetal matter** every day;
- give a vitamin/mineral supplement if necessary.

(d) Administer **drugs and vaccines** as recommended by local specialists.

**Three possible types of feed mixes (in percent of total weight)
for laying chickens**

Ingredients	Types of feed		
	1	2	3
Maize or millet meal	70	58	58
Rice polishings	—	20	—
Brans (wheat, rice)	—	—	22
Wheat middlings	—	—	8
Groundnut cake, milled	17	10	—
Blood meal	10	—	—
Bones, calcinated/ground	2	—	—
Salt	1	—	—
Concentrated mix (proteins/vitamins/minerals)	—	12	12
Total	100	100	100

22. Remember to gradually reduce **the birds' density** as follows:

- **from the sixth to the ninth week**: maximum eight birds/m^2 or at least 0.125 m^2/bird;
- **from the tenth week to first laying**: maximum six birds/m^2 or at least 0.165 m^2/bird;
- **egg-laying period**: maximum four birds/m^2 or at least 0.25 m^2/bird.

23. During the egg-laying period, **manage your chicken house** properl

(a) **Collect the eggs** several times each day:

- morning, midday and evening in cool weather;
- twice in the morning, plus midday and evening in war weather.

(b) **Change the litter** in the nests twice a week.
(c) Clean and refill **drinkers** every morning or more often if necessar
(d) **Disinfect the house** at least once a year.
(e) Keep good **records of egg production** (see Section 166, **Manage ment, 21/2**) and discuss them with a local specialist to check if yo are doing well. During the first year, **the average laying rate** shou be at least **40 percent** (i.e. 40 eggs/100 days) or 140 eggs per he During the second year, this laying rate normally decreases by 20 25 percent.

- **remove and replace hens** that stop laying eggs and ask advic from the local specialist.

24. Usually, laying hens **produce eggs for 12 to 18 months**, and the should be culled when 18 to 24 months old. As soon as the chicke house is empty, clean and disinfect it thoroughly. It is best to leave empty for two months to reduce the risk of disease.

73 Integrated duck rearing

1. The simplest method is to purchase **two-week-old ducklings** from a specialized farm and rear them to:

- either eight- to 12-week-old **table ducks** of 1.7 to 2.5 kg each; or
- five- to six-month-old **laying ducks** to be reared for another 12 to 18 months to produce eggs.

2. Another simple way is to obtain **one-day-old ducklings** from a specialized farm and nurse them for two weeks, before integrating them with your fish.

3. As you have learned for poultry earlier (see Section 72), the **amount of manure** available for your fish ponds will vary greatly during the growing period of the ducklings. It will then stabilize as the ducks reach their adult live weight and begin to lay eggs. Take this into account when deciding which kind of duck to raise, how many to each pond and where to keep them (see below). Consider also the difference in length of the duck-rearing cycles and plan well ahead to replace stocks as often as necessary.

Choosing the duck breed

4. Select the duck breed among those available locally according to the kind of production you plan:

(a) For the **production of table ducks**: you will probably have the choice between a domesticated breed, such as the Peking duck, Muscovy or some hybrid.

(b) For **the production of eggs**: laying breeds such as the Indian Runner, Khaki-Campbell or Bali are usually preferred in rural areas.

5. To have the full benefits of direct fertilization of the pond water by the ducks, it is best to select a breed that likes staying on water for long periods such as the Peking duck. Before the final choice, consult local specialists and other duck farmers.

Nursing one-day-old ducklings

6. Select a well-drained site on which to build **a small brooding unit**, which should include:

- a brooding house;
- an outside pen; and
- an outside splashing pool.

7. There are two ways to make the **brooding area** within the brooding house:

(a) The best way is to make a wooden or metal **frame** with a floor standing about 1 m above the main floor level. Cover it with strong **wire mesh** (2-mm gauge and 2- to 2.5-cm mesh). You will need an **area of 200 to 250 cm^2** per duckling (40 to 50 birds/m^2). Fence this area with 30-cm-high wire mesh.

(b) You can also use **a brooder** similar to the one described earlier for chicks (see Section 72). Use **dry litter**, changed at least every two days. You may use dry grass, maize bran, wood shavings and rice straw cut into 5- to 7-cm pieces, but do not use fine sawdust or dry tree leaves. You will need a fenced area of 400 to 500 cm^2 per duckling (20 to 25 birds/m^2).

8. The **outside pen** should be well fenced against predators and preferably have a concrete floor for easy cleaning. Provide access from the brooding area to the pen through a tight-fitting door. You will need an area of 0.2 m^2 per duckling (five birds/m^2).

9. **The outside splashing pool** should be at most 15 to 20 cm deep and surrounded by flat slopes. It should be lined either with a plastic sheet or, even better, with concrete. Place it in the middle of the pen. A good size is 1 to 1.5 m^2 for 50 ducklings.

10. **Equip the brooding area** as follows:

(a) For **heating** use storm lanterns protected by netting and an oil drum cut in half lengthwise if necessary, as described for chicks (see Section 72). Young ducklings are very susceptible to cold. Adapt them progressively to ambient air temperature by adopting the following schedule.

Progressive adaptation of ducklings to air temperature

Day	1	2	3	4	7	8-14
Air temp. (°C)	35	32	29	26	23	20-18

(b) For **lighting** at night, always leave at least one lamp on so that the ducklings feel secure and keep eating and drinking at will. Keep a spare lamp readily available.

(c) For **drinking**, provide water in a trough covered with wide mesh screen. Change the water often, or better still, provide flowing water.

(d) For **eating**, use an automatic feeder for dry feed mixture or suitable commercial pellets. You may also use a screened trough in which small quantities of feed should be given from five to seven times a day. You will need at least 1 m of feeding space per 14 birds. Provide minced, fresh vegetal matter such as aquatic plants or vegetable leaves daily.

(e) To aid their digestion, add a tray with **2 to 3 mm of gravel**.

(f) As a source of calcium, add a tray with **limestone grit**.

Dry feed mixture for young ducklings

Ingredients	Percentage fresh weight
Maize or millet	59
Oilseed cake	26
Blood meal	10
Bones, calcinated	4
Salt	1

Note: Also add some vitamin/mineral concentrate

11. You may also use chick feed (see Section 72) or, if necessary, a mix of maize meal, breadcrumbs and minced hard-boiled eggs.

12. **Proceed with the nursing** in the following way:

(a) **During the first week**:

● plan ahead for an average feed consumption of maximum 30 g/d/bird or 750 g/d for 25 Peking ducks;

● carefully regulate the air temperature;

● clean the brooding house at least once every two days;

● **if the weather is warm enough** (at least 26°C), **from the fourth day on**, let the ducklings run in the outside pen and use the splashing pool, providing extra feed next to it.

b) **During the second week**:

- plan ahead for an average feed consumption of maximum 56 g/d/bird or 1.4 kg/d for 25 Peking ducks;
- start using moist mash, but do not make it so wet that it becomes sticky;
- regulate the air temperature as far as necessary, especially at night, not letting it drop below 18°C in the brooding house;
- **if weather permits**, release the ducklings daily into the outside pen to splash in the pool and feed next to it, but at night, restrict them to the brooding house;
- if the night air temperature remains above 18 to 20°C, you may even **transfer your ducklings to a small pond**, but you should protect them well from predators and provide a simple shelter (floor space of at least 1 m^2 per 20 birds) against hot sun and cool winds.

3. If your ducklings have been well managed, they should grow quickly and steadily. **Peking ducklings** should on average weigh 125 to 180 g at the end of the first week and 210 to 380 g by the end of the second week.

How to produce table ducks and fish

4. Older ducklings are much more hardy and easier to raise than chickens. They can also feed to a larger extent on **simple feeds** such as grasses, legumes, vegetables, brans and grains, which can all be easily produced on your farm (see, for example, Sections 41, 42 and 45).

15. **To integrate duck and fish production** in a simple way, you have a choice of arrangements:

- **a central platform** from which the ducks roam **over the whole pond**;
- **a dry run** on a pond dike and a wet run **in the whole pond**;
- **a dry run** on the pond dike and a wet run in **part of the pond**. (A shelter or duck house is then usually provided, and the latter may be built either over the pond or on a dike.)

Note: ducks may eat small fish, so do not raise them on breeding ponds or nursery ponds.

16. If you are **using the central platform**, make it preferably of plastic-coated wire mesh (2 × 2 cm mesh) supported by wooden posts in the pond, 30 to 40 cm above the maximum water level. This provides a protected place for the ducks to rest and feed at will. They can easily reach the platform from each side through inclines made of wire mesh or plank ladders floating at one of their ends. A feed trough stands in the middle of the platform. Base the total platform area on an average six to ten ducks per m^2. Ducks are allowed to roam freely over the whole pond area, which has **several advantages**:

- their manure falls directly into the pond and fertilizes it;
- ducks are safe at night from predators and theft;
- they search for food over the entire pond, which provides them free food including proteins and vitamins, thereby reducing feed costs and helping to control weeds, snails, tadpoles, insect larvae, etc.

Such a system, however, has also **some disadvantages**:

- the platform may be expensive to set up and needs good maintenance;
- supplementary duck feed must be taken regularly to the platform;
- the pond dikes can be greatly damaged by the ducks and so should be protected either by a cover (bamboo, brushwood, etc.) or separated from the pond itself by a low fence (split bamboo, wire mesh, etc.).

17. A second choice would be to establish a **fenced dry run** on a pond dike. You will need about 1 to 1.5 m² per duckling. The dike should be well protected to avoid excessive damage. Provide a simple **shelter** (1 m² per 15 ducks) against strong sun and wind. Place a **feed trough** close to the water's edge. Let the ducks roam over the **entire pond area** to feed on natural food. Protect the rest of the dikes as suggested above. The main disadvantage of this system is the loss of about 40 percent of the manure and of the uneaten feed on the dry run.

18. Alternatively, **you may confine the ducklings to part of the pond area**, preferably away from the dikes, by building a small fence from split bamboo, plastic or wire mesh, or wooden poles:

- extend the fence about 40 cm above maximum water level;
- underwater, 60 to 70 cm are sufficient to stop diving ducks;
- use a 4- to 5-cm mesh to allow good water circulation through the fence;
- limit the space available to the ducks for swimming to about 2 to 4 m² per bird.

19. In this case, it is usual to provide the ducklings with **some protection at night** against predators and theft. You may choose from one of the following:

(a) **A well-protected dry run**: Build a good fence around it. Allow about three birds/m². Provide a **simple shelter** (1 m² per 15 birds) and **feed trough**. Protect the access dike. It may be best to pave it and the dry run, for example, with cobble, brick, concrete or soil cement.

(b) **A simple duck house built over the pond**: Use a slatted floor (wood or split bamboo at 1-cm intervals) or a mesh floor (1.25-cm mesh of 8-gauge wire) stretched over a strong frame. In the first case, allow three to four birds/m². Alternatively, allow about five to seven birds/m². Use low (40 to 60 cm) walls (for example, wood off-cuts) with wire mesh (maximum 2.5-cm mesh) above. A thatched roof is good. Provide direct access to the pond with an oblique wooden ladder. Either allow free access to the pond at night for drinking or provide **ample water** in the house for the night. Place a **feed trough** in the middle of the floor.

(c) **A duck house built on a dike**: Use 60-cm high walls made of bricks, cement blocks or wood off-cuts. Above, use wire mesh (maximum 2.5-cm mesh). A concrete floor is preferable. You may either use a **deep litter** (straw, wood shavings, sand) on this floor, allowing three to four birds per m² or build **a mesh floor** about 1 m above the dike, allowing about five to seven birds/m². A thatch or corrugated sheet roof can be used and insulated underneath as necessary. Provide easy and direct access to one or two ponds, and allow free access to the pond at night or provide **ample water** in the duck house. Place a **feed trough** in the middle of the floor.

20. **Feeding your growing ducklings** properly is one of the keys to successful production of table ducks.

21. **If you use an extensive system**, for example by having a low density of three to five ducklings/100 m^2 roaming free over the entire pond area, it will be enough to supplement their natural diet with a 10- to 12-percent protein feed mostly made of grasses, legumes, brans and grains produced on the farm.

22. **If you use a more intensive system**, for example by having a higher density of 15 to 20 ducklings/100 m^2 of total pond area or by confining them to only part of the pond, you will need a better feed containing 15 to 18 percent proteins. (Various possibilities are given in the following chart.)

Higher-quality duck feeds containing 15 to 18 percent protein and suitable for more intensive rearing (in percent of total weight)

Ingredients	Feeds for weeks 3 and 4			Feeds for week 5 and later		
	1*	2	3*	1*	2	3*
Maize or millet meal	59	58.3	—	74	62	50**
Oilseed cake, milled	26	—	—	20	—	—
Blood meal	10	—	—	3	—	—
Bones, calcinated/ground	4	—	—	2	—	—
Salt	1	—	—	1	—	—
Bran, wheat or rice	—	—	30	—	—	—
Chick feed (Section 72)	—	—	70	—	—	—
Broiler feed (Section 72)	—	—	—	—	—	50
Sunflower cake, milled	—	19.6	—	—	19.1	—
Cottonseed cake, milled	—	3	—	—	3	—
Molasses	—	3.7	—	—	3	—
Lime	—	1.8	—	—	2.4	—
Concentrated mix (proteins/vitamins/minerals)	—	13.6	—	—	10.5	—
Total	100	100	100	100	100	100

* To be supplemented by a concentrated mix of vitamins and minerals
** Whole grains

23. Duck feeds may be distributed in different forms:

- commercial **pellets** are best, and reduce waste;
- **dry feed mixtures** result in high waste;
- crumbling but non-sticky **moist feeds** are quite good if freshly prepared.

24. According to the type of feed, there are **two ways to distribute feeds**:

- pellets and dry feeds are best given through **automatic feeders**, which can easily be built locally from wood;
- moist feeds are best given in **troughs**, which are built wider than for broiler chickens (see Section 72) to reduce waste.

25. Build your feeders and feed troughs, and decide how many you need for each batch of ducks, on the following basis:

- three to four weeks old: you require at least 1 m per 11 birds;
- later: you need more space, at least 1 m per eight birds.

26. **Plan your feed requirements for intensive duck rearing** as suggested in the chart on the next page. From six weeks old on, distribute part of the daily food ration early in the morning (60 percent) and part around noon (40 percent). The average **food conversion ratio** over a period of eight weeks should be about 3.5:1.

Feed requirements for intensive rearing of Peking ducks in fish ponds

Duck age (weeks)	Food ration (g/d/bird)	Feed required (kg/week/bird)	For 25 ducklings (kg/week)
3	At will	0.550	13.8
4	At will	0.900	22.5
5	150	1.050	26.3
6	150	1.050	26.3
7	170	1.190	29.8
8	170	1.190	29.8
9 and later	200	1.400	35.0

27. Remember also to provide the following for your ducklings:

- **clean drinking-water**, whenever they do not have access to the pond;
- **fresh vegetal matter**, especially legumes and grasses;
- **small gravel**, 2 to 3 mm in size.

28. If you have taken good care of your ducklings, they should grow well. If you are raising **Peking ducks**, regularly obtain the live weight of a few birds and compare it to the chart on page 227. If you are doing quite well, it should be close to the upper curve. If it is below the lower curve, your ducks are not growing well at all, and you should improve their feed.

Beware: the transfer of young ducklings from the nursing area to the ponds causes great stress and some birds may die. Transfer early in the morning. Be calm and gentle to reduce losses.

Producing duck eggs and fish together

29. If a good market is readily available, you might also consider the production of duck eggs instead of, or in addition to, table ducks. You should then choose a locally available egg-laying breed such as Khaki Campbell.

30. **Khaki-Campbell and Muscovy ducks consume less feed than Peking ducks**; plan your requirements for their intensive rearing as follows.

Feed rations for Khaki-Campbell ducks

Duck age (weeks)	Food ration (g/d/bird)	Consumption (g/week/bird)
0-1	15.7	110
1-2	28.6	200
2-3	42.9	300
3-4	77.1	540
4-5	92.9	650
5-6	100	700
6-7	107	750
7-12	120	840
12-20	135	945

Average growth curves of Peking ducks

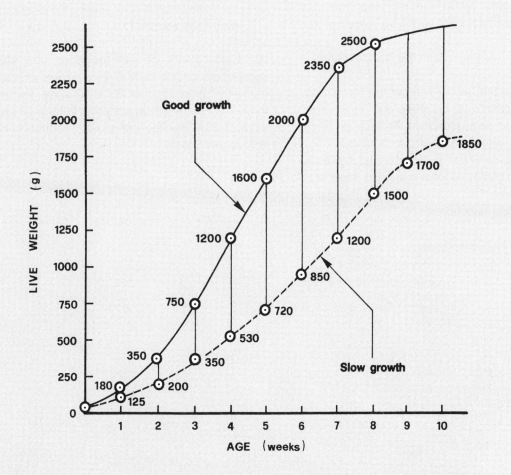

31. The procedure for producing laying ducks is similar to the one described earlier for the production of table ducks using **a duck house** (see paragraph 19), except for the following.

(a) Plan for **more floor space**, at least 0.4 m^2/bird on plain floor and 0.3 m^2/bird on mesh floor, not including drinking, feeding and nesting areas.

(b) Continue rearing the ducklings until they reach **the point of laying** in their fifth or sixth month.

(c) Provide them with **nest boxes**, at the rate of one per five birds. They should each measure 30 cm wide by 40 cm deep and 30 cm high. Add a 15-cm-high front lip. Build them either on the floor or 30 cm off it, against a side wall or a rear wall of the duck house.

(d) **Line the nests** with clean grass or straw, to be changed regularly.

(e) From the day they start laying, give the ducks a good **layer feed** (see chart), at the rate of 150 to 200 g/day/bird. Split this ration equally over two meals, morning and noon.

(f) Make sure that they have **water available** whenever they cannot reach the pond.

(g) Feed plenty of **fresh vegetal matter** daily.

(h) **Collect the eggs** several times a day, clean them if necessary and store them in a cool place. Keep good records (see Section 166 **Management, 21/2**).

32. If your ducks are well managed, they should lay **a large number of eggs** of 50 to 80 g each over a period of about 12 months. On average, this may be as high as 60 to 80 eggs per day per 100 ducks, although there may be large seasonal variations. Then as laying declines, usually when ducks are 18 months old, it is time to **cull them** and replace them with a new batch.

Three types of feeds for egg-laying ducks

Ingredients	Types of feed		
	1	2	3
Broken rice	—	—	50
Rice bran	—	—	30
Whole maize	32.9	30	—
Maize meal	16.8	—	—
Wheat middlings	17.9	—	—
Sunflower cake, milled	13.0	—	—
Molasses	5.0	—	—
Lime (calcium)	8.5	—	—
Concentrated mix (proteins/vitamins/minerals)	5.9	—	—
Layer feed (Section 72)	—	70	20
Total	100.0	100	100

MEASUREMENT UNITS

Length/thickness	**m**	metre	*Water flow*	**l/m**	litre per minute
	cm	centimetre = 0.01 m		**m³/s**	cubic metre per second
	mm	millimetre = 0.001 m		**m³/d**	cubic metre per day
	mil	thousandth of an inch = 0.0254 mm			
			Temperature	**°C**	centigrade, degree Celsius
Area	**m²**	square metre			
	ha	hectare = 10 000 m²	*Alkalinity*	**SBV**	acid-binding capacity unit = 50 mg CaCO₃/l
Volume/capacity	**l**	litre			
	ml	millilitre = 0.001 l	*Proportion/ concentration*		
	m³	cubic metre = 1 000 l		**ppm**	part per million = mg/l = g/m³
	gallon, US	= 3.785 l		**%**	percent
	gallon, Imperial	= 4.546 l			
			Stocking/ application rate	**kg/ha**	= g/10 m²
Weight	**g**	gram		**t/ha**	= 10 kg/100 m² = 100 g/m²
	mg	milligram = 0.001 g			
	kg	kilogram = 1 000 g	*Aeration/ oxygenation*		
	t	tonne = 1 000 kg		**mg/l/h**	rate of change in concentration
				mg/kg/h	rate of specific oxygen consumption
Time	**s**	second		**g/h or kg/h**	rate of oxygen demand, or supply by aerator
	min	minute = 60 s			
	h	hour = 60 min = 3 600 s		**kg/kWh**	aeration efficiency
	d	day = 24 h			

229

COMMON ABBREVIATIONS

Water quality (typical units used)

DO	dissolved oxygen (mg/l)
pH	chemical reaction, acid or alkaline
SD	Secchi disc transparency (cm)
TA	total alkalinity (mg $CaCO_3$/l, SBV)
TSS	total suspended solids (mg/l)
t°	temperature (°C)

Major chemicals (in water, soil, lime and fertilizers)

C	carbon
Ca	calcium
CO_2	carbon dioxide
$CaCO_3$	calcium carbonate
N	nitrogen
NO_3	nitrate (compound)
P	phosphorus
PO_4	phosphate (compound)
K	potassium
K_2O	potash

Fertilization of ponds

N/P	ratio of nitrogen to phosphorus
C/N	ratio of carbon to nitrogen
DM	dry matter

GLOSSARY OF TECHNICAL TERMS[1]

ACTERIA
Very small, one-cell organisms, often developing into large colonies and unable to produce carbon compounds through photosynthesis (see below); primarily responsible for decay/decomposition (see below) of dead plant and animal matter

ENTHOS
Organisms living in and on the mud of the pond bottom

CARBONATES
Acid salts of carbonic acid (see carbonates); their solution in water contains the ion HCO_3 such as calcium bicarbonate $Ca(HCO_3)_2$

ARBONATES
Salts of carbonic acid, a compound formed from carbon dioxide (CO_2) in contact with water; for example calcium carbonate, $CaCO_3$

ELLULOSE
Organic compound that constitutes the essential part of the solid framework of plants; also found in the animal body

OPPICING
Cutting trees close to ground level with the aim of producing shoots from buds near the base of the plant

DECOMPOSITION
The process by which organic compounds decay, disintegrate or rot into simpler chemical compounds or elements

DENSITY DETRITUS
Any disintegrated organic matter accumulated in water, on mud or on soil

FOOD WEB
Pathways through which nutrients added to a pond are converted into fish flesh

FUNGI
Group of plants lacking the ability to produce carbon compounds/organic matter through photosynthesis (see below); include yeasts and moulds

HUMUS
Broken down organic material in organic fertilizers, composts or soils, in which much of the nutrient is available for fertilizing purposes

LIMESTONE
Natural rock consisting chiefly of calcium carbonate

MULCH
Loose covering made of organic residues (e.g. cut grass, straw, tree leaves) applied on the surface of the soil, mainly to conserve moisture and check weed growth

[1] This glossary contains definitions of the technical terms marked with an asterisk (*) the first time they appear in the text.

PERENNIAL — Land vegetation which grows and survives through more than one year, and which is usually in leaf throughout the year

PHOTOSYNTHESIS — Process by which green plants containing chlorophyll convert light energy into chemical energy, producing organic matter from minerals; especially the production of carbon compounds from carbon dioxide (see above) and water, with the release of oxygen

PHYTOPLANKTON — Very small aquatic plants which are suspended in the water; plant component of the plankton

PLANKTON — The various very small organisms, either plants (see phytoplankton) or animals (see zooplankton), which live suspended in the water

POLLARDING — Cutting back the crown of a tree with the object of producing a close head of shoots (a pollard) beyond the reach of browsing animals, for commercial purposes or for amenity

PROTOZOA — Very small single-cell animal organisms, sometimes living in colonies

RESPIRATION — Process by which a living organism, plant or animal, combines oxygen with organic matter, releasing energy, carbon dioxide (see above) and other products

ZOOPLANKTON — Very small aquatic animals which are suspended in the water; animal component of the plankton

ZOOPLANKTERS — Members of the zooplankton (see above) community

FURTHER READING

oyd, C.E. 1982. *Water quality management for pond fish culture.* Amsterdam, Elsevier (Developments in Aquaculture and Fisheries Sciences, 9). 318 p.

O. 1983. Freshwater aquaculture development in China. *FAO Fisheries Technical Paper*, 215. 125 p.

pher, B. and Y. Pruginin. 1981. *Commercial fish farming with special reference to fish culture in Israel.* New York, John Wiley and Sons. 261 p.

et, M. 1986. *Textbook of fish culture. Breeding and cultivation of fish.* Farnham, Surrey, UK, Fishing News Books Ltd., 2nd ed. 438 p.

tle, D. and J.F. Muir. 1987. *A guide to integrated warm-water aquaculture.* Stirling, Scotland, Institute of Aquaculture. 238 p.

Maar, A., M.A.E. Mortimer and I. Van der Lingen. 1966. Fish Culture in Central East Africa. *FAO Fisheries Series*, 20. 160 p.

McLarney, W. 1984. *The freshwater aquaculture book. A handbook for small-scale fish culture in North America.* Points Roberts, Washington, Hartley and Marks, Inc. 583 p.

Seagrave, C., 1988. *Aquatic weed control.* Farnham, Surrey, UK, Fishing News Books Ltd. 154 p.

Stickney, R.R. 1979. *Principles of warmwater aquaculture.* New York, John Wiley and Sons. 375 p.

Woynarovich, E. and L. Horváth. 1980. The artificial propagation of warmwater finfishes. A manual for extension. *FAO Fisheries Technical Paper*, 201. 183 p.

NOTES

NOTES

NOTES

WHERE TO PURCHASE FAO PUBLICATIONS LOCALLY
POINTS DE VENTE DES PUBLICATIONS DE LA FAO
PUNTOS DE VENTA DE PUBLICACIONES DE LA FAO

• **ANGOLA**
Empresa Nacional do Disco e de
Publicações, ENDIPU-U.E.E.
Rua Cirilo da Conceição Silva, N° 7
C.P. N° 1314-C
Luanda

• **ARGENTINA**
Libreria Agropecuaria
Pasteur 743
1028 Buenos Aires
Oficina del Libro Internacional
Alberti 40
1082 Buenos Aires

• **AUSTRALIA**
Hunter Publications
P.O. Box 404
Abbotsford, Vic. 3067

• **AUSTRIA**
Gerold Buch & Co.
Weihburggasse 26
1010 Vienna

• **BANGLADESH**
Association of Development
Agencies in Bangladesh
House No. 1/3, Block F, Lalmatia
Dhaka 1207

• **BELGIQUE**
M.J. De Lannoy
202, avenue du Roi
1060 Bruxelles
CCP 000-0808993-13

• **BOLIVIA**
Los Amigos del Libro
Perú 3712, Casilla 450
Cochabamba;
Mercado 1315, La Paz

• **BOTSWANA**
Botsalo Books (Pty) Ltd
P.O. Box 1532
Gaborone

• **BRAZIL**
Fundação Getúlio Vargas
Praia do Botafogo 190, C.P. 9052
Rio de Janeiro
Núcleo Editora da
Universidade Federal Fluminense
Rua Miguel de Frias 9
Icaraí-Niterói
24 220-000 Rio de Janeiro
Editora da Universidade Federal
do Rio Grande do Sul
Av. João Pessoa 415
Bairro Cidade Baixa
90 040-000 Porto Alegre/RS
Book Master Livraria
Rua do Catete 311 lj. 118/119
20031-001 Catete
Rio de Janeiro

• **CANADA**
Le Diffuseur Gilles Vermette Inc.
C.P. 85, 151, av. de Mortagne
Boucherville, Québec J4B 5E6
UNIPUB
4611/F Assembly Drive
Lanham MD 20706-4391 (USA)
Toll-free 800 233-0504 (Canada)

• **CHILE**
Libreria - Oficina Regional FAO
Calle Bandera 150, 8° Piso
Casilla 10095, Santiago-Centro
Tel. 699 1005
Fax 696 1121/696 1124
Universitaria Textolibros Ltda.
Avda. L. Bernardo O'Higgins 1050
Santiago

• **COLOMBIA**
Banco Ganadero
Revista Carta Ganadera
Carrera 9ª N° 72-21, Piso 5
Bogotá D.E.
Tel. 217 0100

• **CONGO**
Office national des librairies populaires
B.P. 577
Brazzaville

• **COSTA RICA**
Libreria Lehmann S.A.
Av. Central
Apartado 10011
San José

• **CÔTE D'IVOIRE**
CEDA
04 B.P. 541
Abidjan 04

• **CUBA**
Ediciones Cubanas, Empresa de
Comercio Exterior de Publicaciones
Obispo 461, Apartado 605
La Habana

• **CZECH REPUBLIC**
Artia Pegas Press Ltd
Import of Periodicals
Palác Metro, P.O. Box 825
Národní 25, 111 21 Praha 1

• **DENMARK**
Munksgaard, Book and
Subscription Service
P.O. Box 2148
DK 1016 Copenhagen K.
Tel. 4533128570
Fax 4533129387

• **DOMINICAN REPUBLIC**
CUESTA - Centro del libro
Av. 27 de Febrero, esq. A. Lincoln
Centro Comercial Nacional
Apartado 1241
Santo Domingo

• **ECUADOR**
Libri Mundi, Libreria Internacional
Juan León Mera 851
Apartado Postal 3029
Quito

• **EGYPT**
The Middle East Observer
41 Sherif Street
Cairo

• **ESPAÑA**
Mundi Prensa Libros S.A.
Castelló 37
28001 Madrid
Tel. 431 3399
Fax 575 3998
Libreria Agricola
Fernando VI 2
28004 Madrid
Libreria Internacional AEDOS
Consejo de Ciento 391
08009 Barcelona
Tel. 301 8615
Fax 317 0141
Llibreria de la Generalitat
de Catalunya
Rambla dels Estudis 118
(Palau Moja)
08002 Barcelona
Tel. (93) 302 6462
Fax (93) 302 1299

• **FINLAND**
Akateeminen Kirjakauppa
P.O. Box 218
SF-00381 Helsinki

• **FRANCE**
Lavoisier
14, rue de Provigny
94236 Cachan Cedex
Editions A. Pedone
13, rue Soufflot
75005 Paris
Librairie du Commerce International
24, boulevard de l'Hôpital
75005 Paris

• **GERMANY**
Alexander Horn Internationale
Buchhandlung
Kirchgasse 22, Postfach 3340
D-65185 Wiesbaden
Uno Verlag
Poppelsdorfer Allee 55
D-53115 Bonn 1
S. Toeche-Mittler GmbH
Versandbuchhandlung
Hindenburgstrasse 33
D-64295 Darmstadt

• **GHANA**
SEDCO Publishing Ltd
Sedco House, Tabon Street
Off Ring Road Central, North Ridge
P.O. Box 2051
Accra

• **GUYANA**
Guyana National Trading
Corporation Ltd
45-47 Water Street, P.O. Box 308
Georgetown

• **HAÏTI**
Librairie «A la Caravelle»
26, rue Bonne Foi, B.P. 111
Port-au-Prince

• **HONDURAS**
Escuela Agricola Panamericana,
Libreria RTAC
El Zamorano, Apartado 93
Tegucigalpa
Oficina de la Escuela Agricola
Panamericana en Tegucigalpa
Blvd. Morazán, Apts. Glapson
Apartado 93
Tegucigalpa

• **HUNGARY**
Librotrade Kft.
P.O. Box 126
H-1656 Budapest

• **INDIA**
EWP Affiliated East-West Press
PVT, Ltd
G-I/16, Ansari Road, Darya Gany
New Delhi 110 002
Oxford Book and Stationery Co.
Scindia House, New Delhi 110 001;
17 Park Street, Calcutta 700 016
Oxford Subscription Agency
Institute for Development
Education
1 Anasuya Ave., Kilpauk
Madras 600 010
Periodical Expert Book Agency
D-42, Vivek Vihar, Delhi 110095

• **IRAN**
The FAO Bureau, International and
Regional Specialized
Organizations Affairs
Ministry of Agriculture of the Islamic
Republic of Iran
Keshavarz Bld, M.O.A., 17th floor
Teheran

• **IRELAND**
Publications Section
Government Stationery Office
4-5 Harcourt Road
Dublin 2

• **ISRAEL**
R.O.Y. International
P.O. Box 13056
Tel Aviv 61130

• **ITALY**
Libreria Scientifica Dott. Lucio de
Biasio "Aeiou"
Via Coronelli 6
20146 Milano
Libreria Concessionaria Sansoni
S.p.A. "Licosa"
Via Duca di Calabria 1/1
50125 Firenze

FAO Bookshop
Viale delle Terme di Caracalla
00100 Roma
Tel. 52255688
Fax 52255155
E-mail: publications-sales@fao.org

• **JAPAN**
Far Eastern Booksellers
(Kyokuto Shoten Ltd)
12 Kanda-Jimbocho 2 chome
Chiyoda-ku - P.O. Box 72
Tokyo 101-91
Maruzen Company Ltd
P.O. Box 5050
Tokyo International 100-31

• **KENYA**
Text Book Centre Ltd
Kijabe Street, P.O. Box 47540
Nairobi

• **LUXEMBOURG**
M.J. De Lannoy
202, avenue du Roi
1060 Bruxelles (Belgique)

• **MALAYSIA**
Electronic products only:
Southbound
Sendirian Berhad Publishers
9 College Square
01250 Penang

• **MALI**
Librairie Traore
Rue Soundiata Keita X 115
B.P. 3243
Bamako

• **MAROC**
La Librairie Internationale
70 Rue T'ssoule
B.P. 302 (RP)
Rabat
Tel. (07) 75-86-61

• **MEXICO**
Libreria, Universidad Autónoma
de Chapingo
56230 Chapingo
Libros y Editoriales S.A.
Av. Progreso N° 202-1° Piso A
Apdo. Postal 18922, Col. Escandón
11800 México D.F.

• **NETHERLANDS**
Roodveldt Import b.v.
Brouwersgracht 288
1013 HG Amsterdam

• **NEW ZEALAND**
Legislation Services
P.O. Box 12418
Thorndon, Wellington

• **NICARAGUA**
Libreria HISPAMER
Costado Este Univ. Centroamericana
Apdo. Postal A-221
Managua

• **NIGERIA**
University Bookshop (Nigeria) Ltd
University of Ibadan
Ibadan

• **NORWAY**
Narvesen Info Center
Bertrand Narvesens vei 2
P.O. Box 6125, Etterstad
0602 Oslo 6
Tel. (+47) 22-57-33-00
Fax (+47) 22-68-19-01

• **PAKISTAN**
Mirza Book Agency
65 Shahrah-e-Quaid-e-Azam
P.O. Box 729, Lahore 3

• **PARAGUAY**
Libreria INTERCONTINENTAL
Editora e Impresora S.R.L.
Caballero 270 c/Mcal Estigarribia
Asunción

• **PERU**
INDEAR
Jirón Apurimac 375, Casilla 4937
Lima 1

• **PHILIPPINES**
International Booksource Center (Phils)
Room 1703, Cityland 10
Condominium Cor. Ayala Avenue &
H.V. de la Costa Extension
Makati, Metro Manila

• **POLAND**
Ars Polona
Krakowskie Przedmiescie 7
00-950 Warsaw

• **PORTUGAL**
Livraria Portugal,
Dias e Andrade Ltda.
Rua do Carmo 70-74, Apartado 2681
1117 Lisboa Codex

• **SINGAPORE**
Select Books Pte Ltd
03-15 Tanglin Shopping Centre
19 Tanglin Road
Singapore 1024

• **SOMALIA**
"Samater's"
P.O. Box 936
Mogadishu

• **SOUTH AFRICA**
David Philip Publishers (Pty) Ltd
P.O. Box 23408
Claremont 7735
South Africa
Tel. Cape Town (021) 64-4136
Fax Cape Town (021) 64-3358

• **SRI LANKA**
M.D. Gunasena & Co. Ltd
217 Olcott Mawatha, P.O. Box 246
Colombo 11

• **SUISSE**
Buchhandlung und Antiquariat
Heinimann & Co.
Kirchgasse 17
8001 Zurich
UN Bookshop
Palais des Nations
CH-1211 Genève 1
Van Diermen Editions Techniques
ADECO
Case Postale 465
CH-1211 Genève 19

• **SURINAME**
Vaco n.v. in Suriname
Domineestraat 26, P.O. Box 1841
Paramaribo

• **SWEDEN**
Books and documents:
C.E. Fritzes
P.O. Box 16356
103 27 Stockholm
Subscriptions:
Vennergren-Williams AB
P.O. Box 30004
104 25 Stockholm

• **THAILAND**
Suksapan Panit
Mansion 9, Rajdamnern Avenue
Bangkok

• **TOGO**
Librairie du Bon Pasteur
B.P. 1164
Lomé

• **TUNISIE**
Société tunisienne de diffusion
5, avenue de Carthage
Tunis

• **TURKEY**
Kultur Yayiniari is - Turk Ltd Sti.
Ataturk Bulvari N° 191, Kat. 21
Ankara
Bookshops in Istanbul and Izmir

• **UNITED KINGDOM**
HMSO Publications Centre
51 Nine Elms Lane
London SW8 5DR
Tel. (071) 873 9090 (orders)
(071) 873 0011 (inquiries)
Fax (071) 873 8463
and through HMSO Bookshops
Electronic products only:
Microinfo Ltd
P.O. Box 3, Omega Road, Alton
Hampshire GU34 2PG
Tel. (0420) 86848
Fax (0420) 89889

• **URUGUAY**
Libreria Agropecuaria S.R.L.
Buenos Aires 335
Casilla 1755
Montevideo C.P. 11000

• **USA**
Publications:
UNIPUB
4611/F Assembly Drive
Lanham MD 20706-4391
Toll-free 800 274-4888
Fax 301-459-0056
Periodicals:
Ebsco Subscription Services
P.O. Box 1431
Birmingham AL 35201-1431
Tel. (205)991-6600
Telex 78-2661
Fax (205)991-1449
The Faxon Company Inc.
15 Southwest Park
Westwood MA 02090
Tel. 6117-329-3350
Telex 95-1980
Cable FW Faxon Wood

• **VENEZUELA**
Tecni-Ciencia Libros S.A.
Torre Phelps-Mezzanina
Plaza Venezuela
Caracas
Tel. 782 8697/781 9945/781 9954
Tamanaco Libros Técnicos S.R.L.
Centro Comercial Ciudad Tamanaco
Nivel C-2
Caracas
Tel. 261 3344/261 3335/959 0016
Tecni-Ciencia Libros, S.A.
Centro Comercial, Shopping Center
Av. Andrés Eloy, Urb. El Prebo
Valencia, Ed. Caraboba
Tel. 222 724
Fudeco, Libreria
Avenida Libertador-Este
Ed. Fudeco, Apartado 254
Barquisimeto C.P. 3002, Ed. Lara
Tel. (051) 538 022
Fax (051) 544 394
Télex (051) 513 14 FUDEC VC
Fundación La Era Agrícola
Calle 31 Junin Qta
Coromoto 5-49, Apartado 456
Mérida
Libreria FAGRO
Universidad Central de Venezuela (UCV)
Maracay

• **ZIMBABWE**
Grassroots Books
100 Jason Moyo Avenue
P.O. Box A 267, Avondale
Harare;
61a Fort Street
Bulawayo

Other countries / Autres pays / Otros países
Distribution and Sales Section
Publications Division, FAO
Viale delle Terme di Caracalla
00100 Rome, Italy
Tel. (39-6) 52251
Fax (39-6) 52253152
Telex 625852/625853/610181 FAO I
E-mail: publications-sales@fao.org

1/5/96

The following is a list of manuals on aquaculture published in the FAO TRAINING SERIES

"Simple methods for aquaculture" series:

Volume 4 — **Water for freshwater fish culture**
1981. 111 pp. ISBN 92-5-101112

Volume 6 — **Soil and freshwater fish culture**
1986. 174 pp. ISBN 92-5-101355-1

Volume 16/1 — **Topography for freshwater fish culture: topographical tools**
1988. 328 pp. ISBN 92-5-102590-8

Volume 16/2 — **Topography for freshwater fish culture: topographical surveys**
1989. 266 pp. ISBN 92-5-102591-6

Volume 20/1 — **Pond construction for freshwater fish culture: building earthen ponds**
1995. 355 pp. ISBN 92-5-102645-9

Volume 20/2 — **Pond construction for freshwater fish culture: pond-farm structures and layouts**
1992. 214 pp. ISBN 92-5-102872-9

Volume 21/1 — **Management for freshwater fish culture: ponds and water practices**
1996. 233 pp. ISBN 92-5-102873-7

In preparation:

Volume 21/2 — **Management for freshwater fish culture: farms and fish stocks**

Other manuals on acquaculture in the FAO TRAINING SERIES:

Volume 8 — **Common carp 1: mass production of eggs and early fry**
1985. 87 pp. ISBN 92-5-102301-8

Volume 9 — **Common carp 2: mass production of advanced fry and fingerlings in ponds**
1985. 85 pp. ISBN 92-5-102302-6

Volume 19 — **Simple economics and bookkeeping for fish farmers**
1992. 96 pp. ISBN 92-5-103002-2